STEM Education

Global Education in the 21st Century Series

Series Editor

Tasos Barkatsas (*RMIT University, Australia*)

VOLUME 2

The titles published in this series are listed at *brill.com/gecs*

STEM Education

An Emerging Field of Inquiry

Edited by

Tasos Barkatsas, Nicky Carr and Grant Cooper

BRILL
SENSE

LEIDEN | BOSTON

All chapters in this book have undergone peer review.

The Library of Congress Cataloging-in-Publication Data is available online at http://catalog.loc.gov

Typeface for the Latin, Greek, and Cyrillic scripts: "Brill". See and download: brill.com/brill-typeface.

ISSN 2542-9728
ISBN 978-90-04-39138-3 (paperback)
ISBN 978-90-04-39140-6 (hardback)
ISBN 978-90-04-39141-3 (e-book)

This book is printed on acid-free paper and produced in a sustainable manner.

Contents

Acknowledgements

A warm and heartfelt thanks to all the authors and reviewers who have so kindly donated their time. We have a fantastic network of academics with extensive experience in their respective fields; it was a joy to work with all of you, and your contribution to this book has been valuable.

Acknowledgements to all authors, to Dr Claudia Orellana, RMIT University, Australia and to Professors John Malone, Curtin University, Australia, Vasilis Gialamas, National University of Athens and Ngai Ying Wong, The Education University of Hong Kong, for their participation in the peer review process.

A special thank you to Claudia Orellana for her proof-reading of the final drafts and to Associate Professor Linda Hobbs for her thoughtful Epilogue chapter.

Figures and Tables

Figures

Tables

Notes on Contributors

Natalie Banks
is an Assistant Principal and mathematics teacher at Rosebery Middle School, with more than 20 years of experience, in both middle and senior secondary sectors. She is passionate about providing meaningful and targeted teaching to enable all students the opportunity to succeed in their learning and believes that developing respectful relationships and trust with students is key. Natalie became involved in the Reframing Mathematical Futures Project with RMIT in 2013 and has led the implementation of strategies as a whole school approach, helping to close the gaps in students' knowledge of multiplicative thinking.

Anastasios (Tasos) Barkatsas
is a Senior Academic in Mathematics and Statistics Education and a Quantitative Data Analyst at the School of Education, RMIT University, Australia and has published more than 100 refereed journal and conference research papers, chapters and books. Tasos is the also Series Editor of the Brill Publishers Series: 'Global Education in the 21st Century,' an Editorial Board member in a number of international research journals and a reviewer in numerous international research journals and conferences. Tasos is currently co-editing two books, which will be published in 2019 as part of his book series (Volumes 2 and 3). His sole authored book (Volume 4), *Learning Mathematics with Technology: Weaving the eternal braid of attitudes, engagement, confidence, gender and achievement,* will also be published in 2019.

Amanda Berry
is a Professor of STEM Education at Monash University, Australia. As a teacher educator and researcher, Amanda's work focuses on the development of STEM teachers' knowledge and the ways in which that knowledge is shaped and articulated through teacher preparation, beginning teaching and inservice learning. Amanda has published extensively in the above areas, including Handbook chapters, international journals and academic texts. She is current editor of the journal, Studying Teacher Education and Associate Editor of Research in Science Education.

Lisa Borgerding
is an Associate Professor of Science Education and a science teacher educator at Kent State University in Kent, OH (USA). She has taught college science, college science education, high school science, informal elementary and middle school science, and preschool STEAM camp science. Dr. Borgerding's

research focuses on evolution and nature of science education, preservice and inservice science teacher development, and children's science conceptions.

Nicky Carr
has been a teacher educator and researcher for over 15 years, and before that worked with and in the Education system in Victoria in policy and program development. Throughout this time Nicky has had a particular interest in the integration of digital technologies into teaching and learning, particularly examining the factors that shape teachers' decisions to adopt technology or not and the impact of the recently introduced Design and Digital Technologies Curriculum on teaching practices. Nicky is also interested in pre-service teacher education more generally, including their self-efficacy in technology. In a previous career, Nicky prepared a report into the take up of Science and Mathematics Education in Australia in the early 1990s. Her background, and interests in the role of technology and teacher education have recently coalesced to focus on pre-service teachers' conceptualisation and actualisation of STEM Education.

Io Keong Cheong
is a secondary mathematics teacher in Macau's Pooi To Middle School. He is interested in self-regulated learning, use of IT in mathematics learning, and mathematical experiments.

Grant Cooper
is a lecturer in science and STEM education at RMIT University. He is an educator, researcher, learner and maker. At present, his research interests include the examination of emerging STEM education discourses, pre-service teacher preparation of STEM-related literacies/perceptions and how digital technologies such as VR have the potential to transform teaching and learning spaces. Grant's research interests also cover the use of statistical analysis in the field of education.

Marc de Vries
worked as a physics teacher in a secondary school, then did his Ph.D. at Eindhoven University of Technology, worked at the Pedagogical Technical College in Eindhoven as a teacher educator, and later as an assistant professor of philosophy of technology at Eindhoven University of Technology. Currently he is a professor of Science Education and a professor of Philosophy of Technology at Delft University of Technology. Marc's research interests are the nature of technology and technological knowledge and design-based concept learning.

Jennifer Earle
is a PhD candidate in STEM education research at the University of Tasmania. As a graduate of the University of Sydney with Honours in Chemistry, followed by a post-graduate Diploma of Education, Jennifer realised her long-standing passion for teaching science and mathematics in years 7–12. Her teaching career spanned several decades and broadened to include training primary teachers in teaching pedagogy in East Timor to being a Science Curriculum Teacher Leader in Tasmania. Jennifer's experiences as an educational leader fuelled her academic interest in STEM education and raised questions that were not yet answered. Her interest is particularly in the nature of capabilities required by industry into the future and how education can meet these needs.

Noleine Fitzallen
conducts research in statistics education. Her thesis was on student reasoning about covariation when using the exploratory data analysis software, *TinkerPlots*. Prior to embarking on her PhD study, Noleine was investigating the integration of ICT in the mathematics classroom. Her statistics education research focus has now shifted to exploring students' development of understanding of modelling with data when conducting investigations embedded within STEM contexts. Her other research interests include investigating the outcomes for undergraduates delivering STEM outreach programs, the constructive alignment of learning in tertiary education, and the assessment of inquiry-based learning.

Tricia Forrester
is an academic in mathematics education who earned a doctorate from the University of Western Sydney. Her teaching and research have focused on the implementation of inquiry-based approaches to mathematics education and the development of mathematical representations and reasoning. She was the University of Wollongong's Education lead on the OLT project *Inspiring Mathematics and Science in Teacher Education* (IMSITE), which utilised cross-disciplinary approaches to improving mathematics and science teacher education. Currently, her research is focused on improving mathematics education in primary, secondary and tertiary settings by creating mathematics learning spaces that support and encourage active participation and collaboration.

Sharon Fraser
is a lecturer in science education in the School of Education at the University of Tasmania. Sharon began her career as a scientist, and after a number of years working in related fields, she taught secondary science, mathematics

and ICT before proceeding into science and STEM educational research. Her research spans science and mathematics curriculum and pedagogy, both school and higher education, as well as teacher education and the professional learning of teachers in school and university. The theme running throughout her academic career and research is capacity building, which she has enacted through learning and teaching enhancement initiatives and STEM research. Her interest in the latter includes several key areas of science education; including a focus on the increased engagement in science (and mathematics) learning through improved teaching practice, and the investigation of what is knowable in science. Sharon enjoys working with pre-service and in-service teachers who love science, are excited about engaging in STEM (or STEAM) and who appreciate the place of these disciplines in enabling the creation of truly capable learners.

Helen Georgiou
is a science educator who earned a doctorate from the University of Sydney and is an Early Career Researcher specialising in physics education. She coordinates study units in primary and secondary science teacher education. Helen's research focuses on how to describe and develop student understanding in science, particularly in areas which consistently cause problems for students. Her PhD, from The University of Sydney, looked at student understanding of thermodynamics. One aspect of Helen's current research draws from sociology in using Legitimation Code Theory to explore the nature of scientific knowledge in various educational contexts.

Andrew Gilbert
is an Associate Professor of Science Education in the Elementary Education program at George Mason University in Fairfax, VA (USA). Over the last 25 years, he has taught across a wide range of educational contexts across multiple countries. His main research interests center on inquiry-based science instruction, social justice, school-based partnerships and the development of science teacher educators. His most recent research projects have explored wonder as a means to inspire future science teachers and children to engage in science and as a tool to engage in integrated STEM approaches.

Ineke Henze
is a qualitative researcher interested in science teacher professional knowledge and beliefs. Currently, she is an instructor at Delft University of Technology teaching courses on methodology of science teaching, focusing on pre-service science teachers' PCK development during school internships. Prior to completing her PhD, she obtained a Master's in chemistry and a Master's in

education and child studies, all from Leiden University. For about two decades she taught chemistry in secondary education.

Linda Hobbs
is an Associate Professor of Education (Science Education) at Deakin University. She teaches primary science education in the Bachelor of Education Course, and unit chairs a fourth year science education unit that is solely based in schools. She also teaches science communication to science and engineering students. Her research interests include partnerships in primary teacher education, out-of-field teaching in secondary schools, and STEM education. She currently leads a multi-institutional Australian Research Council funded project called Teaching Across Subject Boundaries (TASB) exploring the learning that teachers undergo in their first years of teaching a new subject.

Sarah Howard
is an educational technologist who earned a doctorate from the University of Sydney. Her research focuses on the use of new methodological approaches to explore teacher's technology-related change, specifically technology adoption and integration in learning. Her work takes a particular interest in technology change related to subject areas, school culture and the underlying principles of teaching and learning in those areas. Sarah's research is driven by the idea that helping teachers to change and improve their practice, in ways that are appropriate to their context, make it possible to innovate in learning.

Sylvia Sao Leng Ieong
is an Associate Professor in the Faculty of Education, University of Macau. Before she became an Associate Professor in UM, she taught F5&F6 English in Macao Pui Ching Middle School for 16 years. Currently she is also an adjunct professor of FED, Rector's Office Consultant of UM. Her interests include curriculum studies, translation and SI, and English education. She has authored/co-authored over 100 journal papers, several books and translated works. She has received several awards including The Medal of Merit in Education awarded by the Macao SAR Government.

Chunlian Jiang
is an Assistant Professor in the Faculty of Education, University of Macau. Her research interest includes mathematical problem solving, mathematical problem posing, use of IT in the teaching and learning of mathematics, mathematics teacher education, values in mathematics education, and mathematics Olympiad.

Kathy Jordan
has been a teacher, teacher educator and researcher for over thirty years. She has a range of research interests including educational technologies in school education, teacher use of ICT, the sociology of ICT, literacy and its teaching, teacher education and professional experience. She has published widely in these fields and presented at numerous national and international conferences. Kathy has led research projects around supporting beginning teachers face the challenges of being new to the profession and encouraging systemic change in initial teacher education around ICT. Currently, she is leading a research project around partnerships with schools to improve the classroom readiness of graduates.

Belinda Kennedy
is a research fellow and lecturer in the College of Science, Engineering and Health at RMIT University, Melbourne, Australia. Belinda completed a PhD in Plant Physiology at the University of Technology, Sydney Australia and she holds a Graduate Certificate in Tertiary Teaching and Learning. Belinda has broad teaching experience in the Biological Sciences and has completed a number of STEM teaching initiatives for on-campus and off campus programs for students at year all levels. Her educational research studies in STEM have provided her with a clear understanding of the needs of industry and future students in STEM-related areas.

Zsolt Lavicza
After receiving his degrees in mathematics and physics in Hungary, Zsolt began his postgraduate studies in applied mathematics at the University of Cincinnati. While teaching mathematics in Cincinnati he became interested in researching issues in the teaching and learning mathematics. In particular, he focused on investigating issues in relation to the use of technology in undergraduate mathematics education. Afterwards, both at the Universities of Michigan and Cambridge, he has worked on several research projects examining technology and mathematics teaching in a variety of classroom environments. In addition, Zsolt has greatly contributed to the development of the GeoGebra community and participated in developing research projects on GeoGebra and related technologies worldwide. Currently, Zsolt is a Professor in STEM Education Research Methods at Johannes Kepler University's Linz School of Education. From JKU he is working on numerous research projects worldwide related to technology integration into schools; leading the doctoral programme in STEM Education at JKU; teaching educational research methods worldwide; and coordinates research projects within the International GeoGebra Institute.

Tricia McLaughlin

is currently employed at RMIT University, Australia, on secondment to School of Education. Her background includes lecturing in project/construction management and executive officer, Victorian Parliamentary Inquiry into Building Industry. She worked in the Australian construction industry, held positions on industry skills council and as a Federal Government advisor. She has won several competitive research grants, university and national teaching awards. Her research publications include five books.

Wendy Nielsen

is a science educator who earned a doctorate in science education from the University of British Columbia. Her research focuses on teaching and learning strategies in science, particularly in preservice primary teacher education. She currently leads an ARC Discovery Project examining university science student learning through the creation of digital explanations. She has written widely in STEM-related areas of science teacher education, nature of science, history of science and environmental education. Other research interests include doctoral education and supervising teacher knowledge.

Shalveena Prasad

graduated from the University of the South Pacific and has taught Mathematics in schools in Fiji Islands and Australia. She has been teaching mathematics at Plumpton High School for 10 years. Her passion for improving mathematics outcomes in students encouraged her to explore targeted teaching with her colleagues in the classroom and at numeracy groups. As the Head of the Mathematics Faculty, she is responsible for improving the mathematics teaching and learning at the school. Shalveena took on the RMF specialist teacher role in 2014 and is responsible for introducing a targeted teaching approach to multiplicative thinking and mathematical reasoning as part of the school's commitment to the Reframing Mathematical Futures Project. She is also actively involved in the professional development of teachers in the development of authentic tasks to enhance multiplicative thinking in students.

Theodosia Prodromou

is a Cypriot-Australian mathematician, statistician and mathematics educator, who joined the University of New England in Australia in July 2009 after completing her PhD studies at Warwick university in United kingdom. She taught primary and secondary Mathematics in different countries of Europe, and Australia. She has experience of teaching mathematics education to pre-service teachers and in-service teachers within primary, secondary and post-graduate programs. She is involved in European and International research projects. She is the chair of the GeoGebra institute in Australia. Her interests mostly focus on: the relationship between technology and mathematical thinking; integration of digital technologies in the teaching of mathematics; STEM education; Secondary teachers' professional development; statistics education, statistical literacy, use of Big Data in Educational settings. She is working on numerous research projects worldwide related to technology integration into schools; Big Data and Augmented Reality.

Wee Tiong Seah
is an Associate Professor in Mathematics Education at the Melbourne Graduate School of Education, The University of Melbourne, Australia. Wee Tiong's research expertise is in comparative research, with interests in values/valuing in mathematics education, teacher noticing, as well as immigrant and refugee students and teachers in mathematics pedagogy.

Dianne Siemon
is a Professor of Mathematics Education in the School of Education at RMIT University (Bundoora) where she is involved with the preparation of pre-service teachers and the supervision of higher degree students. Di is currently the Director of the *Reframing Mathematical Futures* project, which is working with 32 secondary schools nationally to develop an evidenced based teaching and learning framework for mathematical reasoning in the middle years. She is also actively involved in the professional development of practicing teachers, particularly in relation to the development of the 'big ideas' in number, the teaching and learning of mathematics in the middle years, and the use of rich assessment tasks to inform teaching. Di has directed a number of other large scale research projects including the *Scaffolding Numeracy in the Middle Years Project* (2003–2006), the *Researching Numeracy Teaching Approaches in Primary Schools Project* (2001–2003), and the *Middle Years Numeracy Research Project* (1999–2001). Di is a past President of the Australian Association of Mathematics Teachers and a life member of the Mathematical Association of Victoria.

Li Ping Thong
is a digital media artist and academic, with extensive experience in practicing, teaching and researching across a myriad range of digital media specialisations. Li Ping is currently the Program Manager and Lecturer for the Bachelor of Design (Digital Media) program at RMIT University, Melbourne, Australia. With over 12 years of internationalised higher education teaching experience across Malaysia, Vietnam and Australia, Li Ping has taught undergraduate courses in different specialisations such as animation, interactive media, user experience (UX) design, user interface (UI) design, mobile app development, serious

games and virtual reality. Her PhD study, successfully completed at the Serious Games Institute (SGI) of Coventry University, UK, investigated the learning effectiveness of digital role-playing games (RPG) to accomplish learning outcomes of digital media education, in which she designed and developed Virtual Designer, a 3d role-playing game that enables students to role-play as design practitioners within a simulated workplace environment. Her research interests include interactive storytelling, serious games, interactive media and virtual reality.

Jan H. van Driel
worked as a teacher of chemistry in a secondary school before doing a PhD at Utrecht University. From 1995–2016, he worked at ICLON – Leiden University Graduate School of Teaching, the Netherlands. In 2006 Jan became a full professor of science education. From 2010–2016 he was the director of ICLON. In 2016, Jan moved to the University of Melbourne as a professor of science education. Jan's research focuses on science teachers' knowledge and beliefs in the context of pre-service education and educational reform.

Tessa E. Vossen
worked as an education material developer after obtaining her MSc in Biology and Science Communication. Since 2014, she is a PhD candidate at the Leiden University Graduate School of Teaching. Tessa's research focuses on the connection between research and design in secondary science education in the Netherlands. Tessa's special interests are the knowledge, beliefs and attitudes that teachers and students hold regarding the topic of her research.

INTRODUCTION

STEM Education: An Emerging Field of Inquiry

Tasos Barkatsas, Nicky Carr and Grant Cooper

The second decade of the 21st century has seen Governments and industry globally intensify their focus on the role of science, technology, engineering and mathematics (STEM) as a vehicle for future economic prosperity. Solutions to many of the challenges faced globally, such as climate change, energy and water sources, food security and so on are argued to require much greater capabilities in science and technology (Kelley & Knowles, 2016). Economic opportunities for new industries that are emerging from technological advances, such as those emerging from the field of artificial intelligence also require greater capabilities in science, mathematics, engineering and technologies (OCS, 2014). In response to such opportunities and challenges, Government policies that position STEM as a critical driver of economic prosperity have burgeoned in recent years, for example in Australian the Office of the Chief Scientist has released a series of four STEM-related policy documents since 2012, including *Science, technology, engineering and mathematics: Australia's future* (OCS, 2014) and *Australia's STEM workforce, Science, Technology, Engineering and Mathematics* (OCS, 2016). Similar policies exist in the United Kingdom, the United Canada, India, Malaysia to name but a few.

Common to all of these policies are consistent messages that STEM related industries are the key to future international competitiveness, productivity and economic prosperity. Another common theme of these policies is that investment must be made in STEM industry and infrastructure, but that crucial to this economic prosperity is a commensurate investment in STEM education to ensure a reliable supply of entrants into the future workforce that is well prepared for the predicted growth in STEM-based industries and professions. These policies have clear implications for the education sector to play a significant role in assisting government and industry to achieve their economic aims.

However, such calls for increased STEM capabilities are taking place at the same time as interest and participation by students in STEM learning is declining, particularly in western countries and more prosperous Asian nations (Bøe, Henriksen, Lyons, & Schreiner, 2011; Kennedy, Lyons, & Quinn, 2014; Thomas & Watters, 2015). Concerns about declining participation in

 | DOI:10.1163/9789004391413_001

STEM-related subjects in schools and STEM-related courses in higher education, combined with increased focus on the economic imperatives of STEM industry, has prompted a related collection of policies released by national authorities globally that call for reforms and investment in new initiatives in STEM-related education, including the Australian *National STEM School Education Strategy* (Education Council, 2015), *STEM in the Education State* (Department of Education and Training, 2016) policy of the Victorian Government in Australia, the *STEM Works* policy of the South Australian Government (Department for Education and Child Development), *Success through STEM* (Department of Education UK, 2011). Education systems at all levels are trying to respond through a range of reforms, initiatives and approaches to the issues relating to STEM education. Education systems, universities and schools are attempting to increase student participation and engagement in STEM related studies, particularly focusing on engaging females in STEM education, increasing teachers' STEM capacity and teaching quality and investing in STEM facilities within schools' systems. Key to these objectives is the building of STEM leadership capacity and capability. An increased focus on facilitating partnerships between education providers and industry is also emerging in policy and in practice (The Australian Industry Group, 2017).

However, things are not as clear-cut as governments or industry might hope them to be. Educators at all levels are grappling with the complexities and issues that are emerging in what is a relatively new, and some might argue, ill-defined field. This volume aims to explore the current debates in STEM education globally, in relation to the current policy focus on STEM education and to how STEM is being conceptualised in different contexts. The volume also provides space for researchers in STEM to present empirical work and to discuss how STEM is being implemented in different educational settings and how implantation is influencing and shaping student learning.

The foremost issue that faces the majority of educators who identify with STEM education is the STEM "identity crisis" (Portz, 2015, p. 1) – what is STEM and what is not STEM? How STEM education is conceptualised is the focus of the first two chapters of this volume. The National Science Foundation in the USA originally coined the term STEM in the 1990s, as it grappled with identifying subject areas in the curriculum that most directly impacted economic development (Portz, 2015). However, as Fraser, Earle and Fitzallen argue in Chapter 1, since then "scholars cannot agree either upon a meaningful definition of STEM or activities that enable the capabilities promised through STEM education." This is apparent in debates about whether the acronym should be STEM or STEAM, with the latter incorporating Arts as an integral discipline. Proponents of the STEAM acronym posit that including the Arts promotes the creativity that many see as integral to STEM education and the sorts of skills in demand by industry. The counter argument to this is that the Arts does not have a monopoly on creativity and that creativity can emerge from multiple disciplines and ways of thinking.

In addition to the importance of strong discipline knowledge in the related, yet often still separate, disciplines of science, mathematics, engineering and technology, some conceptualisations of STEM education include consideration of generic skills of as creativity, problem solving, critical thinking, and communication skills (Fitzallen & Brown, 2017), since these are skills valued by industry. In their chapter, Fraser et al. explore the range of current conceptualisations of STEM education, identifying how structural issues with the nature of school curriculum reinforce a siloed approach to STEM education and the relative dominance of science and mathematics education at the expense of technology and engineering which are more recent additions to the curriculum. Fraser et al. argue that the inter-disciplinary nature of STEM education calls for authentic problem-solving or decision-making scenarios and should involve students taking an inquiry approach that incorporates working in groups, which fosters discussion about the problems and potential solutions. But, as the authors point out, there is a need for strong discipline content knowledge and pedagogical content knowledge. The authors argue that the demands placed on teachers by STEM education to have the knowledge and capacity to apply different pedagogies for the separate disciplines is untenable and provides a significant challenge for capacity building of current teachers to teach within individual disciplines as well as to teach across STEM disciplines.

The theme of multiple conceptualisations of STEM is also discussed in Chapter 2 where authors Van Driel, Vossen, Henze and De Vries identify different conceptualisations of STEM as an umbrella term that refers to one or more of the constituent disciplines, a pseudonym for science education, which is an umbrella term in itself as well as referring to interdisciplinary, multidisciplinary, or integrated approaches. Indeed, authors in this volume demonstrate a range of conceptualisations of STEM education, with some reflecting their commitment to their individual discipline, while others advocate for a more integrative approach. Van Driel et al. debate the various conceptualisations before reporting on a specific approach to STEM Education from the Netherlands where research and design are connected as core practices across STEM disciplines. They describe the role of school-industry partnerships in providing students with opportunities to acquire real world STEM experiences and for teachers to work collaboratively with STEM professionals working in local industries that empowers them to develop and implement a version of STEM education that fits their local context, student population and resources.

The complexities of operationalizing STEM policies that are identified in the first two chapters are put into perspective in Chapter 3, in which the author, Jordan, analyses the discourse surrounding STEM education, focusing on a series of policy statements from the Australian context. Jordan argues that the uncritical positioning of schools as important vehicles for STEM education reforms ignores the complexities that exist in most school contexts and that there is a gap between the political rhetoric of STEM education and the realities of STEM education in schools.

The following five chapters report on a range of STEM education issues, initiatives and projects from different sectors – higher education, secondary schools and primary/elementary schools. In Chapter 4, Cooper and Thong explore the possibilities of immersive virtual reality (VR) in advancing the STEM education agenda and its potential application in the classroom. The authors argue that the reliance on textual representations of concepts in STEM education may be one factor that is contributing to student underperformance in STEM disciplines. They suggest that VR, with its visuospatial representations may be particularly beneficial when representing and learning about STEM-related concepts.

The tensions between an integrated approach to STEM education with an emphasis on process-oriented outcomes and a more discipline-based approach are brought to the fore in Chapter 5, where Siemon, Banks and Prasad make a strong case for an emphasis on deep understanding of STEM discipline knowledge as foundational for STEM education. Drawing on the findings of large scale research projects into school mathematics learning and teaching, the authors argue for greater attention to be paid to improving students' capabilities in multiplicative thinking, as it is key to accessing further mathematics learning. The authors believe that without strong foundational discipline knowledge, students capacity to apply that knowledge in more integrated approaches to STEM is limited and acts to discourage participation on more advanced STEM education experiences. They contend that a quality STEM education incorporates both coherent, well-planned discipline learning as well as opportunities to apply this knowledge in rich, integrated settings that require collaborative endeavour and the exercise of process skills.

One of the most common debates about how STEM education is conceptualised is the debate between STEM and STEAM. Proponents of the latter argue that the arts and arts practices can play an integral role in enhancing the creative, exploratory and entrepreneurial capacities that are increasingly being seen as essential for innovative STEM industries (Colucci-Gray, Burnard, Davies, & Stuart, 2017; Harris & Le Bruin, 2017). Gilbert and Borgerding in Chapter 6 argue that, particularly in early childhood settings, the arts – painting, drawing, modelling, and construction – provides pre-literate children with the ability to represent their burgeoning ideas in concrete ways. The authors make a case that the inclusion of the arts in STEM teaching in the early years allows even young children to demonstrate their capabilities to engage in STEM education.

In Chapter 7, Prodromou and Lavicza theorise the role of inquiry-based learning approaches and mathematical inquiry learning, aimed at providing more authentic ways of fostering middle school students' curiosity, risk-taking and negotiation of statistical meaning, qualities that are increasingly important in a Big Data era. Through examining the experiences of middle school teachers implementing inquiry-based statistics pedagogy in middle school classrooms, the authors investigate the challenges faced by teachers in adopting new practices as they attempt to introduce more authentic learning STEM education experiences.

In an attempt to explore the declining interest and participation by secondary students in STEM disciplines, the authors of Chapter 8 explore the role of values in mathematics education in Macau. Jiang, Seah, Barkatsas, Ieong and Cheong explore the convictions held by students about mathematics, asserting that values affect cognitive functions and affective dispositions and that the interplay between individuals' values within a classroom can shape students' participation in mathematics learning. The authors make a case for applying a conative construct to explore in greater depth the values of students and how these can inform and facilitate STEM education, both in terms of how it might be best learnt and how the content might be best represented.

At this point the focus of the book shifts from schools and the experiences of students to that of teachers. Chapters 9, 10 and 11 pay attention to the challenges and issues being faced in preparing new teachers for classrooms where STEM is emphasised, whilst the final chapter in this volume examines new possibilities for enhancing how higher education teachers approach STEM education.

The authors of Chapter 9, Nielsen, Georgiou, Howard and Forrester, broadly conceive STEM as an integrative view of learning where the individual disciplines provide context for explorations across the fields. Further, they conceive STEM as an opportunity for their primary (elementary) pre-service teachers to consider curriculum integration across disciplinary subject areas. This is particularly salient for future primary school teachers who are responsible for teaching across all the school subjects. The authors, responding to curriculum and other regulatory changes within their educational jurisdiction of NSW, Australia, provide numerous examples of approaches to STEM disciplines that illustrate how pre-service teachers can be positioned to take an integrative approach to STEM education.

Building on the theme of pre-service teachers and STEM education, Cooper and Carr, in Chapter 10, report on a small-scale study into primary/elementary pre-service teachers' conceptualisation of STEM. Using surveys, online posts and pre-service teachers' visual representations of STEM conceptualisation, the findings of this study revealed that pre-service teachers commonly conceptualised STEM education as an integrated approach using problem-based learning and inquiry-related pedagogies, that also incorporated critical and creative thinking and problem solving skills. This study highlights the challenges that some pre-service teachers experience in being able to practice their STEM pedagogical practices during professional experience in schools and the relatively low levels of self-efficacy to teach particular areas of STEM, particularly engineering and digital technologies.

An innovative approach to developing pre-service teachers' knowledge and understanding of contemporary STEM contexts and pedagogies through participation in a STEM mentoring initiative for schoolgirls is reported on in Chapter 11. The initiative reported by Berry, McLaughlin and Cooper in this chapter explores the intersection of pre-service teachers' STEM education and the objective of increasing girls' participation in STEM education. The authors report on pre-service teachers' self-perceptions as emerging STEM educators, their understandings of STEM and developing a STEM pedagogy, their understandings of school girls' interest, engagement and learning in STEM, and the value of the project for teachers in preparation.

In the final chapter in this volume McLaughlin and Kennedy report on another innovative approach to enhancing the capacity of teachers in STEM education to engage in cross-disciplinary activities, this time in the context of higher education. Using the concept of skills ecosystems as an approach to mentoring of leadership projects and the promotion of skill networks, the authors explored how participation in cross-disciplinary activities increased academics' awareness and confidence in STEM cross-disciplinary work as well as their understanding of the value of such cross-disciplinary work for students.

References

Bøe, M. V., Henriksen, E. K., Terry Lyons, T., & Schreiner, C. (2011). Participation in science and technology: Young people's achievement-related choices in late modern societies. *Studies in Science Education, 47*(1). doi:10.1080/03057267.2011.549621 (http://dx.doi.org/10.1080/03057267.2011.549621)

Colucci-Gray, L., Burnard, P., Davies, R., & Stuart Gray, D. (2017). *Reviewing the potential and challenges of developing STEAM education through creative pedagogies for*

21st learning: How can school curricula be broadened towards a more responsive, dynamic and inclusive form or education. London: British Association for Research in Education (BERA) Research Commission.

Department for Education and Child Development. (2016). *STEM learning: Strategy for DECD preschool to year 12, 2017 to 2020*. Retrieved from https://www.education.sa.gov.au/sites/g/files/net691/f/decd-stem-strategy-2016.pdf

Department of Education and Training. (2016). *STEM in the education state*. Retrieved from https://www.education.vic.gov.au/Documents/about/programs/learningdev/vicstem/STEM_EducationState_Plan.pdf

Education Council,. (2015). *National STEM school education strategy: A comprehensive plan for science, technology, engineering and mathematics education in Australia*. Retrieved from http://www.educationcouncil.edu.au/site/DefaultSite/filesystem/documents/National%20STEM%20School%20Education%20Strategy.pdf

Fitzallen, N., & Brown, N. (2017). Outcomes for engineering students delivering a STEM outreach and education programme. *European Journal of Engineering Education, 42*(6), 632–643.

Harris, A., & de Bruin, L. (2017). STEAM education: Fostering creativity in and beyond secondary schools. *Australian Art Education, 38*(1), 54.

Kelley, T. R., & Knowles, J. G. (2016). A conceptual framework for integrated STEM education. *International Journal of STEM Education, 3*(1), 1–11. doi:10.1186/s40594-016-0046-z

Kennedy, J. P., Lyons, T., & Quinn, F. (2014). The continuing decline of science and mathematics enrolments in Australian high schools. *Teaching Science, 60*(2), 34–46.

National Research Council. (2010). *Preparing teachers: Building evidence for sound policy*. Washington, DC: The National Academies Press. Retrieved from https://doi.org/10.17226/12882

National Research Council. (2011). *Successful K-12 STEM education: Identifying effective approaches in science, technology, engineering, and mathematics*. Washington, DC: The National Academies Press. Retrieved from https://doi.org/10.17226/13158

Office of the Chief Scientist. (2013). *Science, technology, engineering and mathematics in the national interest: A strategic approach*. Canberra: Australian Government. Retrieved from http://www.chiefscientist.gov.au/wp-content/uploads/STEMstrategy290713FINALweb.pdf

Office of the Chief Scientist. (2014). *Science, technologyy, engineering and mathematics: Australia's future*. Canberra: Australian Government. Retrieved from http://www.chiefscientist.gov.au/wp-content/uploads/STEM_AustraliasFuture_Sept2014_Web.pdf

Office of the Chief Scientist. (2016). *Australia's STEM workforce, science, technology, engineering and mathematics*. Canberra: Australian Government. Retrieved from http://www.chiefscientist.gov.au/wp-content/uploads/Australias-STEM-workforce_full-report.pdf

Portz, S. (2015, April 29). *The challenges of STEM education.* Presented to the 43rd Space Congress, Cape Canaveral, FL. Retrieved from https://commons.erau.edu/cgi/viewcontent.cgi?article=3410&context=space-congress-proceedings

The Australian Industry Group. (2017). *Strengthening school-industry STEM skills partnerships: Final project report.* Retrieved from https://cdn.aigroup.com.au/Reports/2017/AiGroup_OCS_STEM_Report_2017.pdf

Thomas, B., & Watters, J. (2015). Perspectives on Australian, Indian and Malaysian approaches to STEM education. *International Journal of Educational Development, 2*(1), 28–34. doi:10.5703/1288284314653

CHAPTER 1

What Is in an Acronym? Experiencing STEM Education in Australia

Sharon Fraser, Jennifer Earle and Noleine Fitzallen

Abstract

In Australia, a National STEM School Education Strategy provides guidance for a whole school approach to STEM, going so far as to suggest that that new teaching approaches be implemented and evaluated, and that STEM be prioritised in teacher professional learning. In recent years, some Australian teachers/schools have enacted STEM pedagogies, in the face of a curriculum that remains focussed on the siloed, traditional disciplines of science and mathematics, and more recently digital technologies (implicitly incorporating engineering) but a more consistent and equitable approach is needed. This chapter synthesises what is known about STEM as it pertains to primary and secondary teacher practice, leadership in education, and enactment of the Australian curriculum; and summarises the implications for teacher education and professional practice in the future.

Keywords

STEM education – teacher professional learning – STEAM

1 Introduction

Currently in Australia there is a growing sense of urgency for education reforms that will build science, technology, engineering and mathematics (STEM) capacity in its workforce, as well as STEM literacy in the community (Australian Industry Group [Ai Group], 2013, 2015; Office of the Chief Scientist [OCS], 2013). This persistent national interest in STEM originated in the United States of America (U.S.) from a recognition of the importance of the four disciplines in solving problems in the country through innovation and technological advances. Originally, combining the disciplines in this way was considered to be "a strategic decision made by scientists, technologists, engineers, and mathematicians to combine forces and create a stronger political voice" (STEM

 | DOI:10.1163/9789004391413_002

Task Force, 2014, p. 7). As a consequence, the STEM acronym was created and as a concept, it continued to fuel policy and education reforms in the U.S. throughout the 1960s, 70s and 80s (Mohr-Schroeder, Cavalcanti, & Blyman, 2015).

The imperative that it is in Australia's interest to focus firmly and urgently on STEM emerged from both rhetoric and policies in the U.S., as well as reports from government bodies, both nationally and internationally (OCS, 2013). Producing STEM knowledge and its "sensible application" (OCS, 2014, p. 5) in research and innovation was seen to provide the deep knowledge required to increase Australia's standing in the world and to be key to productivity, growth and higher living standards for all Australians (OCS, 2012). Indeed, the Chief Scientist (2014), who provides high-level independent advice to political leaders on matters relating to science, technology and innovation, recognised a need for developing STEM literacy across the workforce, as well as establishing a reliable pipeline of STEM graduates into the workforce. This called for inspirational teaching and a core STEM education for all students.

Beyond the deep knowledge of one or more of the STEM disciplines, catering to students who like mathematics and science, an education in STEM has been described as fostering creative problem solvers who help shape the future (National Academy of Engineering, 2008). By engaging in rich STEM experiences, students develop a range of generic skills and ways of thinking that enable entrepreneurial behaviours, such as creativity, problem solving, critical thinking, and communication skills (Fitzallen & Brown, 2017). STEM education is conceived of as providing "frameworks in which new problems can be tackled" and "STEM graduates cite higher order skills in research, logical thinking and quantitative analysis as the return on their degrees; alongside the qualities of creativity, open-mindedness, independence and objectivity" (OCS, 2016, p. 2). While these skills are relevant to many occupations, the gap between skills demanded by employers and knowledge generated in education systems is seen to be widening (AI Group, 2015). Since the increasing importance of STEM skills in the workforce has not been matched by participation of students in STEM both in schools and tertiary education, there have been calls for the development of more engaging school curricula and pedagogy to attract students to STEM (AI Group, 2015; Berry, Chalmers, & Chandra, 2012). This has been hampered by an insufficient collective understanding of what is meant by integrated and engaging STEM teaching and learning.

2 STEM: The Acronym

It is well reported that even though STEM is an internationally well-known acronym, its meaningful application in educational contexts is less well understood or enacted. When the movement in the U.S. initiated the focus on STEM, it originally proposed a shared notion through the acronym SMET (coined by the National Science Foundation (NSF) in the 1990s), and later reconfigured it to STEM (attributed to Dr Judith Ramaley, Assistant Director NSF, 2001–2004). Ramaley considered a change to the original acronym was warranted as it better highlighted the connectivity between the disciplines (Chute, 2009), and resonated better verbally than SMET. The changed acronym perhaps also challenged the assumption that by putting science and mathematics together at the beginning of the acronym, it somehow positioned them as the more important of the four disciplines. Of course, while we need science and mathematics to understand the world around us, it is through engineering and technology that we interact with and mould it.

Despite increasing attention being paid to educational reforms in STEM, what actually constitutes STEM remains an issue for scholars, curriculum developers and educators. Even today, more than 40 years after the acronym was conceived, scholars cannot agree either upon a meaningful definition of STEM or activities that enable the capabilities promised through STEM education to be achieved (Breiner, Sheats Harkness, Johnson, & Koehler, 2011; Bybee, 2010; English, 2016; Honey, Pearson, & Schweingruber, 2014; Siekmann & Korbel, 2016). On one hand, Breiner and colleagues concluded that a thorough conceptualisation of STEM is not essential when embarking on any STEM initiative, as one that is specific to the initiative will emerge through working on shared outcomes. On the other hand, Shaughnessy (2013) preferred that initiatives be framed from the outset, summarising that "STEM education refers to solving problems that draw on concepts and procedures from mathematics and science while incorporating the teamwork and design methodology of engineering and using appropriate technology" (2013, p. 324). Siekmann and Korbel (2016) posited that inconsistencies in the definition and enactment of STEM, calls for a deconstruction of the concept to create an alternate and more easily understood grouping of capabilities. At the heart of the term STEM, therefore, is the supposition that there is an inherent connectivity of the four disciplines that is well understood and that should lead to meaningful integrated STEM learning. In actuality, in Australia, anecdotal evidence suggests science and mathematics activities tend to dominate STEM learning.

Seeking to perhaps influence and support STEM education in schools, the Australian Education Council approved the following definition – STEM education is:

> a term used to refer collectively to the teaching of the disciplines within its umbrella – science, technology, engineering and mathematics – and

> also to a cross-disciplinary approach to teaching that increases student interest in STEM-related fields and improves students' problem solving and critical analysis skills. (2015, p. 5)

While this definition provides a level of clarity, it does little to assist teachers to know how to support students to draw on concepts and skills from different disciplines to solve everyday problems (Vasquez, 2014). The multi-faceted nature of the definition provides little guidance for teachers, which makes it unlikely it will influence educational policies, programs, and practices.

In 2010, in Australia, Bybee asserted that it was "time to move beyond the slogan and make STEM literacy for all students an educational priority" (p. 31). Balka (2011) suggested that being STEM literate is having "the ability to identify, apply, and integrate concepts from science, technology, engineering, and mathematics to understand complex problems and to innovate to solve them" (p. 7). Zollman (2012) added that being STEM literate is pivotal to addressing the concerns about economic security and societal demands for advances in science and technology, as it addresses the need for people to become fulfilled, productive, and knowledgeable citizens. Positioning STEM literacy as an end goal of STEM education shifts the conversation from being about how to address the individual disciplines, which is the traditional enactment of the curriculum, to a broader perspective that promises to address the goals of STEM education espoused by the Office of the Chief Scientist and other policy makers, both nationally, and internationally.

3 STEM Education and the Australian Curriculum

Cunningham and Villasenor (2016) conceptualise STEM capabilities into four groupings: occupation or discipline-related technical skills (e.g., coding, design, and construction); socio-emotional skills (e.g., resilience, curiosity, and empathy), advanced cognitive skills (e.g., critical and creative thinking), and foundational literacies needed for everyday life (e.g., numeracy) (p. 20). The totality of these capabilities is epitomised in the General Capabilities dimension of the *Australian Curriculum* (Australian Curriculum, Assessment and Reporting Authority [ACARA], 2018). Australia is one of few countries to explicitly include general capabilities in a national curriculum, and they include: literacy; numeracy; information and communication technology (ICT) capability; critical and creative thinking; personal and social capability; ethical understanding; and, intercultural understanding. The General Capabilities are overarching dimensions of the curriculum and enable connections with the full range of discrete

Australian Curriculum subjects, so that the explicit teaching of these capabilities becomes flexible and dependent on teachers' choice of activities (Marginson, Tytler, Freeman, & Roberts, 2013). However, as Taylor (2016, p. 90) discussed, "designing teaching and learning activities to develop students' higher-order abilities can be daunting" for teachers, and hence remains peripheral to teacher practice. Nevertheless, the development of STEM capabilities that are analogous with the expectations of the Australian Curriculum is desirable and consistent with an integrative approach to STEM education (Fitzallen, 2015).

While socially responsible STEM education, focussed on preparing students for 21st century problems and goals, requires the development of higher order capabilities, the structure of the Australian Curriculum from F-10 and the adherence to traditional approaches (Marginson et al., 2013), act as impediments. The structure of the Australian Curriculum (ACARA, 2018), consisting of separate subjects and strands within them, limits an integrated approach to STEM learning and teaching (Fitzallen, 2015). It follows then that students are also assessed within specific curriculum strands such as science understandings and inquiry skills, while their overall achievement measured against the general capabilities remains peripheral.

While science and mathematics are a clear focus of the curriculum, thereby contributing to the 'S' and 'M' in STEM (Bybee, 2010), engineering and technology are either absent or latecomers to the Australian Curriculum. The 'Technologies' learning area, endorsed in 2015, is comprised of two subjects: *Design and Technologies* and *Digital Technologies* emphasising design thinking and computational thinking, respectively (ACARA, 2018). For curriculum designers, these two subjects may be considered to equate essentially with engineering and information technology, providing the opportunity for the development of strong links with other learning areas, including science and mathematics:

> Thinking and working in the Technologies learning area provides rich opportunities for applying, synthesising and extending learning from other learning areas. It uses, for example, scientific knowledge, language, mathematical concepts, aesthetics and an understanding of sociology and human behaviour. (ACARA, 2012, p. 5)

The 'T' and 'E' in STEM, therefore, support the integration of learning and foster the development of interdisciplinary connections and relationships in a manner aligned with the original intentions proposed by Ramaley (Chute, 2009). The increase in teachers in secondary schools teaching out-of-area and the over-supply of generalist primary teachers in Australia (Weldon, 2015), raises questions about the capacity of some teachers to achieve these ideal outcomes.

4 What Are We Asking of Teachers? The Development of STEM Pedagogical Approaches

The document *Science, Technology, Engineering and Mathematics: Australia's Future* (OCS, 2014), highlighted what is viewed as an abiding need for teachers who can teach STEM explicitly:

> Australia's STEM teachers at all levels, from primary to tertiary, must be equipped to deliver course content with confidence and inspiration, and develop all students to their full potential. Curricula and assessment criteria should prioritise curiosity-driven and problem-based learning of STEM – STEM as it is practised – alongside the subject-specific knowledge that STEM requires. (p. 21)

As stated, teachers should be able to "deliver course content with confidence and inspiration" across multiples disciplines at all levels of schooling. Hence, they need a deep and abiding understanding of the content and nature of each of the individual disciplines that contribute to STEM, as well as the manner in which they meaningfully connect in solving real world problems. In secondary education in particular, being proficient in one, or perhaps two of the STEM disciplines might be expected but moving beyond this to teaching through integrated STEM requires individuals to develop more extensive capabilities. They must also make explicit and significant connections between the four disciplines while balancing the need to integrate context and engage and motivate students (Dare, Ellis, & Roehrig, 2018). The extent to which such outcomes can be achieved through the activities of individual teachers, or whether collaborative, team teaching, or indeed a mixture of both approaches is more effective, is yet to be determined. In any case, to achieve such integration, teachers require more targeted professional learning and interventions, supported through extensive and meaningful research and evaluation.

Research in STEM education often neglects the multifaceted aspects of STEM and the need to ensure both the place of STEM literacy and STEM specialisation. While studies report upon integrative approaches taken to improve students' interest and learning in STEM, research at the school level often puts an emphasis on exploring one of the disciplines situated within STEM contexts, rather than exploring the collective integrated learning. For example, when English, Hudson, and Dawes (2013) explored the application of the engineering design process when constructing catapults with year 7 and 9 students, little attention was given to the science, technology, and mathematics outcomes from the activity. Similarly, Fitzallen, Wright, Watson, and Duncan

(2016) only described students' conceptions of heat transfer as they conducted a science experiment, even though the opportunity to explore students' understanding of the application of various materials for insulation and other specific purposes (technology) was embedded in the activity. While Bottge, Grant, Stephens, and Rueda (2010) focused exclusively on the outcomes of the learning and teaching of mathematics situated within a technology learning context.

All of these studies acknowledged the importance of STEM education but none report on the inter-disciplinary nature of the learning, enhanced through the STEM activities. From a research perspective, targeting particular aspects of a learning situation to answer specific questions is not unusual; however, it adds little to evidence-based pedagogical practices suitable for promoting STEM education, that go beyond traditional teaching practices that treat the disciplines as separate entities. While it could be argued that it is not necessary to have integrated learning, it seems such an approach is needed to address the multiple demands placed on STEM education (Bybee, 2010; OCS, 2016; Zollman, 2012).

In general, integrated STEM learning involves real-life contexts and the development of proto-types or models to simulate authentic problem-solving or decision-making scenarios (e.g., English et al., 2013; Moore, Guzey, & Brown, 2014; Ward, Lyden, Fitzallen, & León de la Barra, 2015). They also involve students taking an inquiry approach that incorporates working in groups, which fosters discussion about the problems and potential solutions throughout (Kelley & Knowles, 2016). Moore and Smith (2014) contended that STEM integration can occur from two perspectives – context integration and/or content integration. Context integration refers to the use of some motivator in order to teach some disciplinary content, which is usually mathematics or science. Content integration, however, involves purposefully targeting the various ways of working in STEM – inquiry-based, team-based, or project-based (Kelley & Knowles, 2016) – and disciplinary content as learning goals (Lyden, Ward, Fitzallen, & Panton, 2018). More evidence is needed, and advice provided to teachers, on the most effective pedagogical approaches for facilitating both content and context integration, and the learning potential offered by both methods of STEM integration.

Although an integrated approach to STEM learning is lauded, most pedagogical models focus on one of the four disciplines. Guidance is provided, for example, around the implementation of inquiry-based learning in science through the *Biological Sciences Curriculum Study 5E Instructional Model* (Bybee, 2015). Also, a pedagogical approach appropriate for accommodating the engineering design process is process-orientated guided inquiry learning

(POGIL) (Moog & Spencer, 2008). In statistics education, Watson, Fitzallen, Fielding-Wells, and Madden (2018) advocated for students working through the stages of statistical modelling, termed the Practice of Statistics, within STEM contexts, and project-based learning is the focus for learning about technology (Akgun, 2013). All of these are student-centred, constructivist pedagogies where the students are guided through content focused on one of the STEM disciplines, with active engagement in the learning process. Another issue that cannot be ignored, however, is the relative value of the individual disciplines within integrated STEM, and the need for explicit instruction for some or all students, through the deliberate teaching of mathematics (Fitzallen, 2015), science concepts and skills (Marginson et al., 2013), and raising an awareness of engineering design activities (Marginson et al., 2013). The demands for teachers to have the knowledge and capacity to apply different pedagogies for the separate disciplines is untenable. This is exacerbated by teachers from different STEM disciplines having different perceptions about STEM integration, which lead to different classroom practices (Wang, Moore, Roehrig, & Park, 2011) and by default, inconsistent learning outcomes.

5 What Does This Mean for Teacher Education and Teaching in Schools?

Attracting and retaining teachers in science and mathematics has been a focus for the Australian government for some years (OCS, 2012) and remains an ongoing challenge (Weldon, 2015). While a recent publication by Prinsley and Johnston (2015) outlined steps necessary to recruit and support suitable teachers to enable the effective teaching of science, technology and mathematics, and implement STEM education in primary schools, ensuring there are sufficient STEM teachers in secondary schools remains a challenge. As is common around the world, teachers teaching science, mathematics and technology subjects as out-of-field teachers (Hobbs, 2013; Weldon, 2015) remains high, with 20–30% of mathematics and science teachers having received no teaching methodology education in these fields during their teacher education studies (Beswick, Fraser, & Crowley, 2016).

There is increasing evidence that a teacher's capacity to teach STEM disciplines is linked to their knowledge and academic background in these disciplines (National Research Council [NRC], 2011), with further evidence emerging in regards the impact of teachers with strong mathematical backgrounds on improved learning outcomes of their students (NRC, 2010; Tatto & Senk, 2011). Such evidence has provided the basis for policy change in Australia, in response to the report from the Teacher Education Ministerial Advisory Group (TEMAG): *Action now: Classroom Ready Teachers* (2014). Specifically, changes were made to the program standards (Program Standard 4.4) for primary teacher education courses, to ensure that initial teacher education (ITE) primary students graduate with both general, and specific teaching skills through undertaking a specialisation to enable a deeper focus in a particular subject area. While the *Australian Professional Standards for Teachers* (Australian Institute for Teaching and School Leadership [AITSL], 2017) does not prioritise any particular subject area, mathematics and science are clearly priority areas in certain states, as evidenced by the incentives in place for teacher retraining (e.g., Teach NSW: STEM scholarships for science and mathematics).

Building the capacity of secondary teachers to teach individual STEM disciplines, as well as being capable of teaching through more integrated STEM pedagogies, requires a different approach, one which clearly targets innovative, interdisciplinary pedagogies. While STEM integration is currently at the fore of curriculum and pedagogical initiatives, integration has been interpreted variously across a spectrum of increasing levels: from disciplinary, multidisciplinary, interdisciplinary to transdisciplinary (Honey et al., 2014; Vasquez, Sneider, & Comer, 2013) and evidence for achievement of desired learning has not yet been researched sufficiently (English, 2016). For STEM education to have the desired impact upon student learning and the development of 21st century capabilities, teacher education programs and professional learning of in-service teachers needs to embrace these interdisciplinary pedagogies.

6 Building Teacher Capacity in STEM Pedagogies: Professional Learning

Authentic STEM teaching requires the adoption and adaptation of new pedagogies, which asks much of both initial teacher education courses, and any teacher retraining and/or STEM professional learning programs. Various Australian states provide guidance and professional learning in integrated STEM education, through initiatives that unearth and support varying beliefs about STEM teaching and challenge them through project-based strategies or engineering design processes with an intentional interdisciplinary approach. Most STEM professional learning for teachers is, unfortunately, "... often short, fragmented, ineffective, and not designed to address the specific need of individual teachers" (Wilson, 2011). Such one-off experiences go little way to developing teacher understandings, or challenging beliefs about STEM, or building their confidence or capacity to engage in STEM pedagogies.

One project trialled in Geelong in Victoria (2015–2017) brought together government, industry, university, and school partnerships in the *Successful Students – STEM in Schools* program (http://www.successfulstudents-stem.org.au/stem-in-schools). The intervention sought to raise STEM teaching capability of teachers of Years 7 and 8 science and mathematics teachers through a longitudinal professional learning program. The project leaders recognised and accommodated varying teacher beliefs and existing pedagogical practices and acknowledged that STEM implementation differs from one school to the next. They supported schools to engage with and develop STEM curricula in a manner suitable to their context and capacity, resulting in the identification of five models of STEM collaboration and/or integration used in different contexts (Table 1.1).

TABLE 1.1 The range of collaboration and integration models used by teachers in 2016 of the Successful Students STEM program (from Hobbs, Cripps Clark & Plant, 2017, p. 144, reproduced with permission, http://www.successfulstudents-stem.org.au/stem-in-schools)

Model	Model description and example
S-T-E-M	1. **Teach each discipline separately:** In science classes, there is a renewed focus on using representations to enhance concept development.
SteM	2. **Teach all four but more emphasis on one or two:** A teacher integrates mathematics and science through a challenge-based unit of work where students design an object or device, say a vehicle.
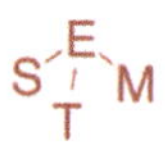	3. **Integrate one into the other three being taught separately:** The engineering processes of team work, identify and investigate a problem, design a solution, and testing and evaluation is added into some science and mathematics units, but there are limited links across the science and mathematics subjects.
STEM	4. **Total integration of all by a teacher:** Science teacher integrating, T, E and M into science. A school introduces a new STEM elective focusing on designing digital solutions to real world problems.
	5. **Divide a STEM curriculum into the separate subjects:** Technology, science and maths teachers design a combined unit and each teacher teaches different components of the unit in their separate subject, and with clear contributions from science, maths and technology subjects in solving a common problem.

The five STEM models of practice identified in the Successful Students STEM program, represent a move from a disciplinary approach to authentic, interdisciplinarity and collaborative curriculum design and pedagogies. However, only two of the models do not privilege at least one discipline over another. Models 4 and 5 represent an "approach to learning that removes the traditional barriers separating the four disciplines … and integrates them into real-world, rigorous, and relevant learning experiences for students" (Vasquez et al., 2013, p. 4). Although the other three models required teachers to support their students to engage in real-world problems that required solutions that draw upon the four STEM disciplines, the approaches perpetuate, to varying degrees, the siloed approach to teaching the four disciplines. By doing so, teachers were supported to move away from working as individuals to collaborating in either subject-based or interdisciplinary teams. This is not an inconsiderable task for all concerned and relied upon buy-in from schools through both structural and instructional leadership, and sufficient resourcing (government funding; university-led intervention). This level of support is not afforded to all schools and making integrated STEM learning a reality for all students is going to take substantial and ongoing resources and commitment from governments, education systems, industry, and above all, teachers.

7 Leveraging Resources: Working with STEM Professionals and the Community

For many years there has been support for Australian students who are interested in individual STEM subjects and want to specialise in one or more STEM disciplines; teachers have enrolled their students in science and mathematics competitions (e.g., Tournament of the Minds; Australian Science Olympiad; Australian Mathematics competitions), and the Commonwealth Scientific and Industrial Research Organisation (CSIRO) has maintained connections with students interested in science (e.g., Double Helix magazine). Most of these types of activities were originally designed for the motivated and 'able' mathematics and science students, rather than with the aim of engaging more students in STEM subjects, and lesser still, enabling the development of STEM literacies.

More recently, however, the science, mathematics, and increasingly STEM footprint in/with schools has increased, with examples including Australian Science Innovations (https://www.asi.edu.au/about/australian-science-innovations/), F1 in schools (https://rea.org.au/f1-in-schools/) and CSIRO STEM Professionals in schools (https://www.csiro.au/en/Education/Programs/

STEM-Professionals-in-Schools). All of these activities/programs link teachers with resources and expertise to engage learners in relevant and authentic STEM experiences, which seek to engage all students not just the already engaged. Unfortunately, their reach is limited, often due to cost both in terms of resources and manpower. Also, some programs are not accessed easily in rural and regional areas. The inability to offer comparable learning experiences to all students raises issues of equity and challenges the ability to enhance STEM education for all.

While a number of these STEM initiatives are relatively recent, research into the outcomes of earlier interventions suggest the numbers of students studying individual STEM subjects in schools, particularly physics and mathematics, has remained static (Kaspura, 2017), regardless of the resources being made available to support STEM education in schools. In fact, longitudinal research conducted in the United Kingdom reported that STEM "enrichment and enhancement activities" that aim to improve understanding and enjoyment in science and encourage long term participation in STEM, even though enjoyable, had no impact upon either increasing or widening STEM participation (Banerjee, 2017).

The extent to which the structure of schools and schooling act as impediments to authentic and engaging STEM education, and hence widening participation, is unknown. In Australia, however, it is common practice in secondary education to timetable discreet lessons in each of the subject/discipline areas in years 7–12. Discounting for the moment, the number of teachers teaching out-of-field, each subject is usually staffed by individual teachers, skilled in specific areas of discipline expertise and pedagogical content knowledge (PCK) (e.g., science) rather than STEM education, per se. For real impact to be felt, therefore, changes to the curriculum to reflect STEM more broadly would be beneficial, alongside skilled teachers in both primary and secondary schools to teach it and committed principals and department heads to lead it.

8 Building Teacher Capacity in STEM Pedagogies: The Role of Principals

School principals are essential to leading the move to improving the specific instruction of STEM subjects (Community for Advancing Discovery Research in Education, 2012) and STEM-focussed learning in their schools. In doing so, they must maintain their responsiveness and resilience in the face of every future initiative in school education. While in a STEM-rich learning environment, students are expected to engage with and learn through STEM activities and their teachers to facilitate such learning, someone needs to create this context, and lead, manage and nurture it. Principals are critical to fostering a shared commitment to developing a STEM culture and improving STEM learning outcomes. The extent to which STEM is prioritised and/or integrated into the school (K-12) curriculum varies, depending upon the school, its context and leadership. Principals are key to developing a school culture and instigating any changes to practice.

The Australian Professional Principal Standard (AISTL, 2017) guides principals as to how they might be effective leaders, naming up three leadership requirements (vision and values; knowledge and understanding; personal qualities, social and interpersonal skills), which are enacted in five areas of professional practice in order to ensure high quality learning and teaching. It is considered the role of principals to lead teaching and learning; improvement, innovation and change; school management; community engagement; and professional development to impact student learning (Louis, Leithwood, Wahlstrom, & Anderson, 2010; Robinson, 2007). In regards the important role principals play in establishing STEM as a strength in schools, they set the tone and drive the culture (OCS, 2016). They are responsible for creating and communicating a vision for STEM and developing and sustaining a culture of effective whole-of-school STEM teaching and learning. Principals are responsible for supporting their staff to build their capacity to teach STEM by providing appropriate resources and effective continuing professional learning (Lai, 2015), which ultimately has the potential to lead to increases in students' success in STEM (Arshavsky, Edmunds, Miller, & Corritore, 2014).

Principals who evaluate the effectiveness and impact of the school's approach to STEM teaching and learning, through generating and acting upon data assist in the development of understanding (Lingwood & Sorenson, 2014) and continual increase of STEM engagement and improved learning outcomes. As Vitoria, Gill, and Mireles (2016) have attested from the outcomes of the STEM Future Action Research Project, "... the expectations of school environments set by school principals and teachers filter into expectations for college readiness and can have positive impacts on student success" (p. 66).

Nelson and Sassi (2005) also identified the importance of principals' content knowledge and beliefs; for example, beliefs about the nature of mathematics and mathematics learning and teaching, as these beliefs affect the ways in which they undertake their leadership roles. This is not to say that all principals need to be able to teach individual STEM subjects or indeed be able to teach STEM in an integrated manner. Principals need to know enough about STEM and its teaching and learning to be able to create a STEM culture in their school and ensure that their staff have the capacity to teach STEM. Exactly what aspects of STEM teaching and learning, resourcing and management are required of principals to lead STEM in schools are yet to be identified. While principals are not required to be STEM experts, they do need to know something about individual subject knowledge, how the curriculum areas can be taught both separately and together, and how STEM is best experienced and learnt by students.

Such baseline understandings about STEM, combined with the more generic leadership abilities outlined in the Professional Principal Standard (AITSL, 2017), are capabilities that principals need in order to lead a whole of school approach to STEM learning. This is a large expectation of principals, and to date there are few resources developed specifically to support them in developing these capabilities. The outcomes envisaged in the numerous reports from the OCS (e.g., 2014, 2016) of more students interested in and engaging in STEM subjects and contributing to the future STEM workforce, will not be realised without principals who can drive a culture of integrated STEM in their schools, and who compile and leverage sufficient and contextually relevant STEM resources for their teachers.

9 What about Other Disciplines – Bringing the A into STEM?

While understandings of STEM remain fragmented, debate continues about its future as a concept useful in guiding education. In 2014, the OCS claimed that the extent to which STEM can operate effectively in the workplace relies upon the deep understandings afforded by the social sciences and humanities, and the social context within which STEM operates. Some researchers (e.g., Edwards, 2010; Taylor, 2016) and reports (e.g., Deloitte, 2015; White, 2010) have advocated for the inclusion of the Arts in the STEM concept (STEAM), arguing that it enriches and broadens STEM education, encouraging students/workers to "create a vision of reimagined work" (Deloitte, 2015, p. 126). Research has shown that its inclusion enhances the perceived degree of problem-solving skills (Cobb Payton, White, & Mullins, 2017) and suggests that there are synergistic and reciprocal benefits between STEM and the Arts, as Guyotte, Sochacka, Costantino, Kellam, and Walther (2015, p. 31) explained:

> While STEAM education might not be the answer to the incessant call for educational reform, there is much to be gleaned from the experiences of our students in such contexts where inquiry, exploration, collaboration, empathy and creativity are brought to the forefront of the curriculum; and where students are nudged into the vibrant and evocative spaces that lie between themselves and the other.

Taylor (2015) argued that STEAM education enables students to engage in transformative learning through being exposed to five interconnected ways of knowing: cultural self-knowing; relational knowing; critical knowing; visionary and ethical knowing; and knowing in action, all of which are essential for addressing key global issues, such the human impact on the climate and the environment (Crutzen & Stoermer, 2000). Siekmann and Korbel (2016) further contended that the value in a holistic and foundational education that prepares students for further education and the workplace may be obscured if schooling focusses on STEM alone. Interconnectivity as a concept underpins the ways scholars think about both STEM and STEAM education. The extent to which teachers have either engaged in such critical reflection of their practice or recognise what embracing either STEM or STEAM requires of them in their teaching practice is still emerging in the literature.

10 Where to from Here?

Both government and industry rhetoric positions STEM expertise and STEM literacy as essential for the future of individual children, and the world they live in, due to industry's dependence on such mastery. How the development of such capabilities is facilitated and enabled remains contested. As summarised in Table 1.1, the Successful Students – STEM Program identified that schools participate in various types of STEM activities/study, which incorporate interdisciplinary perspectives to a greater or lesser extent, along a continuum of possibilities. There is little research evidence supporting the efficacy or influence of any one of these models in increasing students' STEM literacy or engaging students in specialist STEM study in later years of school and post-compulsory education. Yet, the juggernaut requiring the upskilling of teachers and their students for STEM appears unstoppable and is becoming more complex with its conceptual basis being extended to include perspectives of the Arts.

An interdisciplinary STEAM education is more than achieving an increased understanding of its individual disciplines, rather it challenges teacher beliefs about their discipline or disciplines, how they interact, and are taught and learned and therefore, the pedagogies and practices that they adopt. What constitutes useful knowledge in each of the constituent disciplines, and how that knowledge is created, learned and applied, is also challenged by the STEM agenda. For authentic STEM education, a deep and abiding understanding of both the nature and content of each of the disciplines, science, technology, engineering (design technologies) and mathematics is required in order to understand how they are connected and to make those connections explicit for students. Whether STEM is prioritised in the curriculum or a STEAM approach is adopted, changes to the way in which teachers are educated (ITE courses) and the way they are supported (curriculum; school leadership and structures) and sustained (professional learning) need to be made.

Initial teacher education courses offered through Australian universities continue in the main, to emphasise the need for pre-service teachers to develop expertise that mirrors the Australian Curriculum (ACARA, 2018) and aligns with Australian Professional Standards for Teachers (AITSL, 2017). The balanced teaching of discipline knowledge and skills alongside innovative, interdisciplinary STEM pedagogies in these courses is still rare. Entrenched traditional pedagogies remain unchallenged, and therefore the dominant experience of future Australian school teachers. ITE students graduate and enter the profession, therefore, prepared to replicate and perpetuate dominant pedagogical approaches for subject-based education, rather than to engage in and lead STEM-rich education. Universities more generally, have an important role to play in influencing the uptake of interdisciplinary learning in schools. The entry requirements for university courses (e.g., Bachelor of Science), for example, are content-based. A student's achievement in individual subjects (e.g., Chemistry; advanced Mathematics) through high-stakes assessment, are favoured in the university entrance pre-requisites for science and engineering degrees. Thus, the cycle of privileging the single S., T., E., and M. approaches to learning in schools is preserved.

The traditional discipline siloed approach to teaching in schools and the school structures which influence these practices, would benefit from a change. In schools, maintaining the traditional emphasis of subject-centred teaching through rigid school timetabling, for example, not only impedes the collaboration of teachers across subject areas, but it also highlights to students that knowledge comes in chunks, labelled up as subjects. STEM pedagogies are considered inherently interdisciplinary, and while the Australian Curriculum (ACARA, 2018) provides an opportunity to use the General Capabilities as a way of engaging teachers in interdisciplinary discussions and pedagogical approaches, subject-centred school structures work against such collaboration. For teachers, it is the subject-specific curriculum strands that remain the focus of their teaching and are assessed explicitly. Challenging and disrupting such taken-for-granted pedagogical understandings and practices require structural and instructional leadership.

Principals are key to bringing about a change in the culture of the profession and teacher practice, both within their own schools and beyond. They can resource interdisciplinary STEM practices in their schools, critique and change school structures and processes which work against such approaches, and support their teachers to develop their capacity to teach in new ways. They have a role to play in influencing the content that underpins the STEM activities undertaken in their school context, and to oversee the creation and maintenance of appropriate links with local industry. Principals face the dilemma of how best to develop and nourish a STEM-rich school, while still enabling students' engagement with the other curriculum areas. Is this where considering STEAM as an alternative framework for learning contributes to the agenda – as a way of totally rethinking the curriculum? As members of powerful state-based and national bodies, principals have the potential to influence systemic change in support of STEM education, by influencing educational systems to challenge the siloed nature of the national and state-based curricula, and the manner in which learning outcomes are defined and assessed.

STEM education requires high quality purposeful STEM interventions that are evaluated rigorously for impact over the short, medium and long-term. These kinds of data would enable comparisons between models of practice and their impact, linking wherever possible cause to effect. Models of leadership in STEM, structural impediments and enablers to STEM teaching and learning and effective (appropriate and timely) professional learning (including mentoring) for teachers, principals and instructional leaders should be evaluated. The impact of both formal and informal STEM approaches upon student learning and engagement in STEM and STEM disciplines, needs to be investigated as well as their contribution to the STEM education pipeline and workforce.

The debate about STEM 'versus' STEAM education as the way to progress quality learning outcomes in schools remains ongoing. To date, much energy has been invested in arguing the case for and against the inclusion of 'Arts' in the STEM agenda. By addressing student learning through a capability-driven approach, and evaluating rigorously its impact, we believe, makes the inclusion or otherwise of A in the acronym moot. Engaging students in authentic learning that they find personally meaningful, and that enables them to develop important 21st century skills incorporating discipline-related technical skills, socio-emotional skills, critical and creative thinking is paramount. Importantly, optimising students' educational opportunities should be the focus of education, regardless of the outcome of the debate associated with the STEM acronym.

References

Akgun, O. E. (2013). Technology in STEM project-based learning. In R. M. Capraro, M. M. Capraro, & J. R. Morgan (Eds.), *STEM project-based learning: An integrated Science, Technology, Engineering and Mathematics (STEM) approach* (2nd ed.). Rotterdam, The Netherlands: Sense Publishers.

Arshavsky, N., Edmunds, J. A., Miller, L. C., & Corritore, M. (2014). Success in the college preparatory mathematics pipeline: The role of policies and practices employed by three high school reform models. *School Effectiveness and School Improvement, 25*(4), 531–554.

Australian Curriculum, Assessment and Reporting Authority. (2012). *The shape of the Australian curriculum: Technologies.* Retrieved from http://www.acara.edu.au/curriculum/learning-areas-subjects/technologies

Australian Curriculum, Assessment and Reporting Authority. (2018). *The Australian curriculum.* Retrieved January 8, 2018, from https://www.australiancurriculum.edu.au

Australian Industry Group. (2013). *Lifting our Science, Technology, Engineering and Maths (STEM) skills.* Sydney: Author.

Australian Industry Group. (2015). *Progressing STEM skills in Australia.* Sydney: Author.

Australian Institute for Teaching and School Leadership. (2017). *Australian professional standards for teachers.* Retrieved from https://www.aitsl.edu.au/teach/standards

Balka, D. (2011). Standards of mathematical practice and STEM. In *Math-science connector newsletter.* Stillwater, OK: School Science and Mathematics Association. Retrieved from http://ssma.play-cello.com/wp-content/uploads/2016/02/MathScienceConnector-summer2011.pdf

Banerjee, P. A. (2017). Does continued participation in STEM enrichment and enhancement activities affect school maths attainment? *Oxford Review of Education, 43*(1), 1–18.

Berry, M., Chalmers, C., & Chandra, V. (2012, November 24–27). *STEM futures and practice: Can we teach STEM in a more meaningful and integrated way?* Instructional Innovations and Interdisciplinary Research in STEM Education, 2nd International STEM in Education Conference, Beijing. Retrieved from http://stem2012.bnu.edu.cn/data/long%20paper/stem2012_82.pdf

Beswick, K., Fraser, S. P., & Crowley, S. (2016). No wonder out-of-field teachers struggle! Unpacking the thinking of expert teachers. *Australian Mathematics Teacher, 72*(4), 16–21.

Bottge, B. A., Grant, T. S., Stephens, A. C., & Rueda, E. (2010). Advancing the math skills of middle school students in technology education classrooms. *NASSP Bulletin, 94*(2), 81–106.

Breiner, J. M., Sheats Harkness, S., Johnson, C. C., & Koehler, C. M. (2011). What is STEM? A discussion about conceptions of STEM in education and partnerships. *School Science and Mathematics, 112*(1), 3–11.

Bybee, R. W. (2010). Advancing STEM education: A 2020 vision. *Technology & Engineering Teacher, 70*(1), 30–35.

Bybee, R. W. (2015). *The BSCS 5E instructional model: Creating teachable moments.* National Arlington, TX: Science Teachers Association Press.

Chute, E. (2009, February 10). STEM education is branching out. *Pittsburgh Post-Gazette.* Retrieved from http://www.post-gazette.com/news/education/2009/02/10/STEM-education-is-branching-out/stories/200902100165

Cobb Payton, F., White, A., & Mullins, T. (2017). STEM majors, art thinkers (STEM – Arts): Issues of duality, rigor and inclusion. *Journal of STEM Education, 18*(3), 5. Retrieved from https://jstem.org/index.php?journal=JSTEM&page=article&op=view&path%5B%5D=2101

Community for Advancing Discovery Research in Education. (2012). *STEM smart brief: Preparing and supporting STEM educators*. Retrieved from http://cadrek12.org/sites/default/files/Preparing%20Supporting%20STEM%20Educators_FINAL.pdf

Crutzen, P. J., & Stoermer, E. (2000). The anthropocene. *Global Change Newsletter, 41*, 17–18.

Cunningham, W., & Villasenor, P. (2016). *Employer voices, employer demands, and implications for public skills development policy connecting the labor and education sectors.* World Bank Policy Research Working Paper, No. 6853. Washington, DC: World Bank. doi: 10.1596/1813-9450-6853.

Dare, E. A., Ellis, J. A., & Roehrig, G. H. (2018). Understanding science teachers' implementations of integrated STEM curricular units through phenomenological multiple case study. *International Journal of STEM Education, 5*(1), 4. doi:10.1186/s40594-018-0101-z

Deloitte. (2015). *Tech trends 2015: The fusion of business and IT.* Retrieved from http://www2.deloitte.com/au/en/pages/technology/articles/tech-trends-2015.html

Education Council. (2015). *National STEM school education strategy: A comprehensive plan for Science, Technology, Engineering and Mathematics education in Australia.* Canberra: Author.

Edwards, D. (2010). *Artscience: Creativity in the post google generation.* Cambridge, MA: Harvard University Press.

English, L. D. (2016). STEM education K-12: Perspectives on integration. *International Journal of STEM Education, 3*(3), 1–8.

English, L. D., Hudson, P. B., & Dawes, L. (2013). Engineering-based problem solving in the middle school: Design and construction with simple machines. *Journal of Pre-College Engineering Education Research, 3*(2), 43–55.

Fitzallen, N. (2015, June 28–July 2). STEM education: What does mathematics have to offer? In M. Marshman (Eds.), *Mathematics education in the margins: Proceedings of the 38th annual conference of the Mathematics Education Research Group of Australasia, Sunshine Coast* (pp. 237–244). Sydney: Mathematics Education Research Group of Australasia. Retrieved from http://www.merga.net.au/documents/RP2015-22.pdf

Fitzallen, N., & Brown, N. (2017). Outcomes for engineering students delivering a STEM outreach and education programme. *European Journal of Engineering Education, 42*(6), 632–643.

Fitzallen, N., Wright, S., Watson, J., & Duncan, B. (2016, November 27–December 1). Year 3 students' conceptions of heat transfer. In M. Baguley (Ed.), *Transforming educational research: Proceedings of the Australian Association for Research in*

Education conference. Melbourne: Australian Association for Research in Education. Retrieved from http://www.aare.edu.au/publications-database.php/10672/year-3-students-conceptions-of-heat-transfer

Guyotte, K. W., Sochacka, N. W., Costantino, T. E., Kellam, N. N., & Walther, J. (2015). Collaborative creativity in STEAM: Narratives of art education students' experiences in transdisciplinary spaces. *International Journal of Education & the Arts, 16*(15), 1–38. Retrieved from http://www.ijea.org/v16n15/

Hobbs, L. (2013). Teaching out of field as a boundary crossing event: Factors shaping teacher identity. *International Journal of Science and Mathematics Education, 11*(2), 271–297. Retrieved from http://dx.doi.org.simsrad.net.ocs.mq.edu.au/10.1007/s10763-012-9333-4

Hobbs, L., Cripps Clark, J., & Plant, B. (2017). Successful Students STEM Program: Teacher learning through a multifacted vision for STEM education. In R. Jorgensen & K. Larkin (Eds.), *STEM education in the junior secondary* (pp. 133–168). Singapore: Springer.

Honey, M., Pearson, G., & Schweingruber, H. (2014). *STEM integration in K-12 education: Status, prospects, and an agenda for research*. Washington, DC: National Academies Press.

Kaspura, A. (2017). *Engineers make things happen*. Canberra: Engineers Australia.

Kelley, T. R., & Knowles, J. G. (2016). A conceptual framework for STEM education. *International Journal of STEM Education, 3*(1), 1–11. doi:10.1186/s40594-016-0046-z

Lai, E. (2015). Enacting principal leadership: Exploiting situated possibilities to build school capacity for change. *Research Papers in Education, 30*(1), 70–94. Retrieved from http://dx.doi.org/10.1080/02671522.2014.880939

Lingwood, S. A., & Sorenson, J. B. (2014). Paper copters and potential: Leveraging afterschool youth development trainers to extend the reach of STEM programs. *Afterschool Matters, 20*, 39–46.

Louis, K. S., Leithwood, K., Wahlstrom, K. L., & Anderson, S. E. (2010). *Investigating the links to improved student learning: Final report of research findings*. Minneapolis, MN: Center for Applied Research and Educational Improvement.

Lyden, S., Ward, L., Fitzallen, N., & Panton, L. (2018, November 21–23). *Exploring a STEM education pedagogy: Teachers' perceptions of the benefits of an extended integrative stem learning program*. Paper presented at 5th International STEM in Education Conference, Brisbane.

Marginson, S., Tytler, R., Freeman, B., & Roberts, K. (2013). *STEM: Country comparisons*. Melbourne: Australian Council of Learned Academies.

Mohr-Schroeder, J., Cavalcanti, M., & Blyman, K. (2015). STEM education: Understanding the changing landscape. In A. Sahin (Ed.), *A practice-based model of STEM teaching: STEM Students On the Stage (SOS)™* (pp. 3–14). Rotterdam, The Netherlands: Sense Publishers.

Moog, R. S., & Spencer, J. N. (2008). POGIL: An overview. In R. S. Moog & J. N. Spencer (Eds.), *Process Oriented Guided Inquiry Learning (POGIL): ACS symposium series*

(Vol. 994, pp. 1–13). Washington, DC: American Chemical Society. Retrieved from http://pubs.acs.org/isbn/9780841274235

Moore, T. J., Guzey, S. S., & Brown, A. (2014). Greenhouse design to increase habitable land: An engineering unit. *Science Scope, 37*(7), 51–57.

Moore, T. J., & Smith, K. A. (2014). Advancing the state of the art of STEM integration. *Journal of STEM Education, 15*(1), 5–10.

National Academy of Engineering. (2008). *Changing the conversation: Messages for improving public understanding of engineering*. Washington, DC: The National Academies Press. Retrieved from https://doi.org/10.17226/12187

National Research Council. (2010). *Preparing teachers: Building evidence for sound policy*. Washington, DC: The National Academies Press. Retrieved from https://doi.org/10.17226/12882

National Research Council. (2011). *Successful K-12 STEM education: Identifying effective approaches in Science, Technology, Engineering, and Mathematics*. Washington, DC: The National Academies Press. Retrieved from https://doi.org/10.17226/13158

Nelson, B. S., & Sassi, A. (2005). *The effective principal: Instructional leadership for high-quality learning*. New York, NY: Teachers College Press.

Office of the Chief Scientist. (2012). *Mathematics, engineering and science in the national interest*. Canberra: Austrailian Government. Retrieved from http://www.chiefscientist.gov.au/wpcontent/uploads/STEMstrategy290713FINALweb.pdf

Office of the Chief Scientist. (2013). *Science, technology, engineering and mathematics in the national interest: A strategic approach*. Canberra: Australian Government. Retrieved from http://www.chiefscientist.gov.au/wp-content/uploads/STEMstrategy290713FINALweb.pdf

Office of the Chief Scientist. (2014). *Science, technologyy, engineering and mathematics: Australia's future*. Canberra: Australian Government. Retrieved from http://www.chiefscientist.gov.au/wp-content/uploads/STEM_AustraliasFuture_Sept2014_Web.pdf

Office of the Chief Scientist. (2016). *Australia's STEM workforce, Science, Technology, Engineering and Mathematics*. Canberra: Australian Government. Retrieved from http://www.chiefscientist.gov.au/wp-content/uploads/Australias-STEM-workforce_full-report.pdf

Prinsley, R., & Johnston, E. (2015). *Transforming STEM teaching in Australian primary schools: Everybody's business*. Canberra: Office of the Chief Scientist. Retrieved from http://www.chiefscientist.gov.au/wp-content/uploads/Transforming-STEM-teaching_FINAL.pdf

Robinson, V. M. J. (2007). *School leadership and student outcomes: Identifying what works and why*. Winmalee: Australian Council for Educational Leaders.

Shaughnessy, M. (2013). Mathematics in a STEM context. *Mathematics Teaching in the Middle School, 18*(6), 324.

Siekmann, G., & Korbel, K. (2016). *Defining 'STEM' skills: Review and synthesis of the literature.* Adelaide: National Centre for Vocational Education Research. Retrieved from https://files.eric.ed.gov/fulltext/ED570655.pdf

STEM Task Force. (2014). *Innovate: A blueprint for Science, Technology, Engineering, and Mathematics in California public education.* Dublin, CA: Californians Dedicated to Education Foundation. Retrieved from https://www.cde.ca.gov/pd/ca/sc/documents/innovate.pdf

Tatto, M. T., & Senk, S. (2011). The mathematics education of future primary and secondary teachers: Methods and findings from the teacher education and development study in mathematics. *Journal of Teacher Education, 62*(2), 121–137.

Taylor, P. C. (2015). Transformative science education. In R. Gunstone (Ed.), *Encyclopedia of science education* (pp. 1079–1082). Dordrecht: Springer.

Taylor, P. C. (2016, August 7–9). *Why is a STEAM curriculum perspective crucial to the 21st century?* In Proceedings of the Australian Council for Educational Research (ACER) Research Conference 2016, Improving STEM Learning, What will it Take? Brisbane.

Teacher Education Ministerial Advisory Group. (2014). *Action now: Classroom ready teachers.* Retrieved from https://docs.education.gov.au/system/files/doc/other/action_now_classroom_ready_teachers_print.pdf

Vasquez, J. A. (2014). STEM: Beyond the acronym. *Educational Leadership, 72*(4), 10.

Vasquez, J. A., Sneider, C., & Comer, M. (2013). *STEM lesson essentials, grades 3–8: Integrating Science, Technology, Engineering, and Mathematics.* Portsmouth, NH: Heinemann.

Vitoria, M., Gill, P. S., & Mireles, S. V. (2016). Principal preparation in STEM: An action research project. *The Journal of the Effective Schools Project, 23*, 62–68.

Wang, H. H, Moore, T. J., Roehrig, G. H., & Park, M. S. (2011). STEM integration: Teacher perceptions and practice. *Journal of Pre-College Engineering Education Research, 1*(2), 2. Retrieved from https://doi.org/10.5703/1288284314636

Ward, L., Lyden, S., Fitzallen, N., & León de la Barra, B. (2015). Using engineering activities to engage middle school students in physics and biology. *Australasian Journal of Engineering Education, 20*(2), 145–156.

Watson, J., Fitzallen, N., Fielding-Wells, J., & Madden, S. (2018). The practice of statistics. In D. Ben Zvi, J. Garfield, & K. Makar (Eds.), *International handbook of research in statistics education* (pp. 105–137). Cham: Springer.

Weldon, P. R. (2015). *The teacher workforce in Australia: Supply, demand and data issues* (Policy Insights No. 2). Camberwell: Australian Council for Educational Research.

White, H. P. (2010). *STEAM not STEM whitepaper: An agreement on what drives the US economy in the future.* Retrieved from http://steam-notstem.com

Wilson, S. M. (2011, May 10–12). *Effective STEM teacher preparation, induction, and professional development.* In the National Research Council's Workshop on Successful STEM Education in K-12 Schools, Washington, DC. Retrieved from http://sites.nationalacademies.org/cs/groups/dbassesite/documents/webpage/dbasse_072640.pdf

Zollman, A. (2012). Learning for STEM literacy: STEM literacy for learning. *School Science and Mathematics, 112*, 12–19.

CHAPTER 2

Delivering STEM Education through School-Industry Partnerships: A Focus on Research and Design

Jan H. van Driel, Tessa E. Vossen, Ineke Henze and Marc J. de Vries

Abstract

This chapter describes an approach to STEM education that focuses on connecting research and design as core practices across the STEM disciplines. In this approach, school-industry partnerships provide students with opportunities to acquire real world STEM experiences. Collaboration between teachers, within and across schools, and with STEM professionals working in local industries are an essential element in the implementation of this innovation. Consequently, schools and teachers are empowered to develop and implement a version of STEM education that fits their local context, student population and resources. Research is needed to investigate the impact of this approach on the attitudes and behaviours of students, teachers and STEM professionals.

Keywords

Attitudes – collaboration – design – partnerships – research

1 Introduction

The acronym STEM (Science, Technology, Engineering and Mathematics) has gained substantial traction in education around the globe in the last decade, and is associated with the expectation that STEM education will boost student interest and achievement, and that pursuing STEM studies will enhance students' employability. However, international research (Marginson, Tytler, Freeman, & Roberts, 2013) has demonstrated that the STEM subjects are often taught in ways that fail to engage young people. Traditional teaching methods and 19th century content such as 'acids and bases' dominate classrooms, making it hard for students to recognise the relevance of STEM for their daily lives

 | DOI:10.1163/9789004391413_003

and their futures. Moreover, STEM subjects are often perceived as difficult, requiring extra effort from students to achieve. And why would students work hard for something that does not make much sense to them?

This lack of engagement with STEM subjects is an issue in many countries. For example, a recent report on the situation in Australia (Kaspura, 2017) revealed that the percentage of female students in Year 12 who study physics or advanced mathematics is only 6 per cent (for male students, these percentages are 21 and 11.5, respectively). Worryingly, these percentages have dropped considerably, for both males and females, since the mid-1990s (Kennedy, Lyons, & Quinn, 2014). To counter these trends, there is now a global call for innovation in STEM education, and many initiatives are underway. These initiatives have a huge variation in focus, scale and duration. Some focus on increasing STEM participation by girls, others focus on specific areas, such as coding. In Australia, a web portal (STAR portal, 2017) was recently launched, containing over 250 initiatives. To date, there is not much research available to demonstrate the impact on student learning and interest in STEM.

In this chapter, we will focus on an innovation in STEM education in the Netherlands, in which *Research and Design* is implemented as a project-based subject, focused on real world problems. This innovation aims to increase student engagement by providing them with authentic STEM experiences through school-industry partnerships. We will discuss research and design as interdependent core practices across the STEM disciplines and present some findings of studies on the attitudes of teachers and students towards these practices. This leads into an exploration of the affordances and requirements associated with this approach to STEM education, focusing on collaboration between schools and industries. Our aim is to identify characteristics of STEM education that engages students, teachers and STEM professionals.

2 What Is STEM Education?

Interpretations of STEM education differ across the world. In some cases, STEM education is used as an umbrella term, to refer to the teaching of one or more of the constituent disciplines. Sometimes STEM education is used as a pseudonym for science education, which is an umbrella term in itself. However, STEM education can also refer to interdisciplinary, multidisciplinary, or integrated approaches. Educational policies like the Next Generation Science Standards (NGSS Lead States, 2013) place emphasis on providing stronger connections between STEM disciplines, because "most global challenges concerning energy, health, and the environment (e.g., climate change, sustainability) require an interdisciplinary (and frequently, international) perspective involving math, science, and technology" (Shernoff, Sinha, Bressler, & Ginsburg, 2017, p. 2). Also, STEM education may be used as a label for approaches that emphasise the development of certain skills, such as problem solving or critical thinking. Bybee (2010) describes STEM literacy as "the conceptual understandings and procedural skills and abilities for individuals to address STEM-related personal, social, and global issues," placing emphasis on both conceptual knowledge and procedural skills, like inquiry. Policy documents often emphasise increasing student knowledge about career opportunities in STEM (NRC, 2012). Finally, STEM education is sometimes used to advocate for the introduction of contemporary content and technology in the curriculum: robotics, virtual reality, 3D printers, and more.

The above indicates that there are several problems with the term STEM education:

- In most curricula around the world, 'STEM' is not a subject or a course. Instead, schools offer mathematics, biology, chemistry, physics, and (sometimes) integrated science. Moreover, engineering rarely features in curricula for primary and secondary education, whereas technology is typically limited to information or digital technologies.
- The disciplines under the STEM umbrella each have their own concepts and principles, and their teaching approaches can be quite different. Bringing them together under one label could suggest that the STEM disciplines are (more or less) 'the same thing,' which then might promote a range of misconceptions, such as 'if you can teach one, you can teach them all.' The problems associated with teaching out-of-field have been well documented, in particular, for the teaching of mathematics (Hobbs, 2013; Ingersoll, 1998).
- An emphasis on the development of 'STEM skills,' such as collaboration and critical thinking, may lead to undervaluing the importance of developing a deep understanding of disciplinary knowledge, and create challenges for teachers in finding an appropriate balance between teaching knowledge and skills.

Obviously, there are also many problems with school subjects being taught in isolation from each other. Although the integrity of the disciplines needs to be respected, the relationships between them are important and need to be made more visible and explicit, both at the level of the curriculum and the classroom. This requires teachers to learn about the central concepts and principles of each other's subjects and to work together across the STEM disciplines to develop ways to teach and assess problem solving and inquiry skills coherently and consistently across subjects and school years. We would recommend limiting the use of the term STEM education to curricula or subjects that have been purposefully developed to connect the STEM disciplines, for example, through an interdisciplinary or integrated approach. A useful definition has been proposed by Johnson (2013):

> Integrated STEM education is an instructional approach, which integrates the teaching of science and mathematics through the infusion of the practices of scientific inquiry, technological and engineering design, mathematical analysis and 21st century interdisciplinary themes and skills. (p. 367)

In the next sections, we will focus on a particular approach to STEM education, in which research and design are the practices that students engage in to address real world problems, thus developing interdisciplinary knowledge and skills.

3 Research & Design

In this chapter, we argue that research and design are the core of the work of 21st century STEM professionals. Research is mostly associated with science and mathematics and is driven by questions in search for new knowledge. In other words, at the start of a research activity, the answer is unknown. Design, commonly seen as the work of engineers and technicians, is characterised by seeking optimal solutions for a real-life problem. In design, there is never *one* answer. In current STEM practice, professionals with different disciplinary backgrounds work together in teams, and research and design are typically interlinked in their projects. Note how different this is from the way STEM subjects are usually taught in primary and secondary schools, where individual students are usually preoccupied with finding the one, correct answer to problems imposed on them.

Research and design can be seen as two separate practices with distinct goals and histories (Williams, Eames, Hume, & Lockley, 2012). Research is typically employed to explain, explore or compare certain phenomena or events by collecting and analysing data (Creswell, 2008). Design is commonly aimed at developing or improving products or services (de Vries, 2005). However, research and design have several characteristics in common: both are concerned with challenging, ill-structured problems or questions (Hathcock, Dickerson, Eckhoff, & Katsioloudis, 2015), and both are iterative practices. Different models of the research process have been published (e.g., Kolodner, Gray, & Fasse, 2003; Willison, & O'Regan, 2008), however, this process is generally described in terms of the following components: orientation on a research question; generate hypotheses; plan research; collect data; organise and analyse data; conclude and discuss; communicate and present. The design process generally consists of the following phases (Kolodner et al., 2003; Mehalik, Doppelt, & Schuun, 2008): clarify problem; assemble program of requirements; plan design; construct prototype; test prototype; repeat steps to optimise prototype; analyse product; communicate and present.

Several authors have demonstrated the affordances of connecting research and design (Kolodner et al., 2003; Mehalik et al., 2008). For instance, design projects often include research activities. Design activities, on the other hand, can be incorporated in a research project, for example, when designing a device to perform measurements, or when designing experiments (Fallman, 2003). Kolodner et al. (2003) visualised this relationship between investigation (research) and design within STEM education (Figure 2.1). They propose a learning-by-design approach consisting of activities specific to investigating and designing. Whenever there is a 'need to know' during the design cycle, an investigation, or research, needs to be conducted, in which students acquire knowledge they need to complete the design challenge. Similarly, the 'need to do' that may arise during an investigation, refers to the design (for instance, of a tool) that is needed to continue the research.

In current educational policy documents in the USA, like the NRC Framework (NRC, 2012) and the Next Generation of Science Standards (NGSS Lead States, 2013), research and design activities are mentioned as important focal points in K-12 science and engineering education. In these documents, science and engineering practices do not have their own separate process descriptions but have similar phases. However, the documents distinguish between science and engineering as two different practices with different goals: answering questions for science and solving problems for engineering.

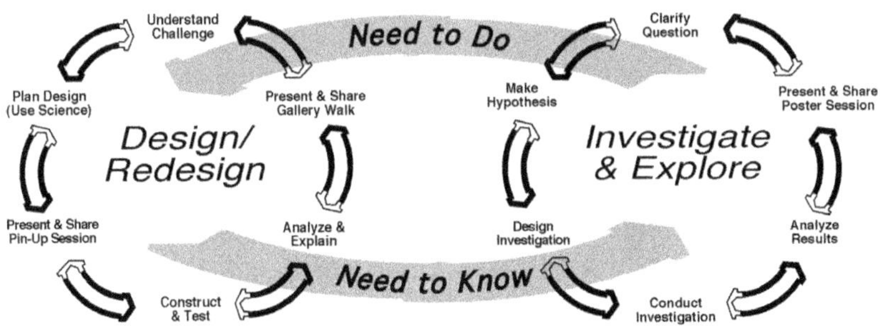

FIGURE 2.1 The interconnected cycles of design and research (from Kolodner, Gray, & Fasse, 2003, reprinted with permission)

4 Research and Design: An Innovation in Dutch STEM Education

Common in many STEM education initiatives around the world is the idea that students can be engaged by working on authentic problems, and that this engagement leads to the development of knowledge and skills. This idea underpins a successful initiative in the Netherlands, called *Technasium*. The core of this innovation is the implementation of a new subject from Years 7 to 12, called *Research & Design* (R&D), alongside the regular school subjects. This subject is taught 4–6 periods per week, requiring schools to create space in the timetable, for instance, by redistributing curriculum time across STEM and art subjects. R&D is entirely project based. The projects are negotiated with parties external to the school, such as local industries and businesses. Students work in teams of 3 to 5 on these projects. Initiated by a couple of parents in 2004 in one school, this approach has gained a lot of traction, with almost 100 schools, operating in regional networks of 4 or 5, having adopted it and being certified to call themselves Technasium. These schools are required to have a studio or workplace, where students can work on their projects, however, facilities and resources differ between schools. Local industries have committed to working with these schools on an ongoing basis, providing input and expertise for the projects and supervision for groups of students, who do their projects mostly at school and partly on-site. Groups of students are also supervised by a school teacher. A training and certification scheme for teachers, recognised by the government, supports the implementation of the program.

The main aims of *Research & Design* have been documented in a national curriculum document. These aims focus on (1) acquainting students with STEM professions, and (2) letting students work on up-to-date and authentic STEM questions, in order to stimulate them to develop skills as competent researchers and designers (Bruning & Michels, 2014). Projects are typically initiated by industries who act as 'clients,' providing students with real research and design problems. For example, in one project a local company asked students to optimise an algae reactor, with a list of factors that influence algae growth, and a plan for upscaling the company's reactor. Whereas specific content knowledge is provided by the 'client,' the role of the R&D teacher from the school is that of a coach who helps students to develop skills like planning, teamwork and perseverance.

In summary, *Research & Design* is an example of integrated STEM education, that uses teaching approaches that differ from those that are common in science and mathematics education. Although R&D has not yet been extensively researched, some research is underway to explore R&D students' attitudes towards research and design (Vossen, Henze, Rippe, van Driel, & de Vries, 2018).

4.1 *Challenges in Teaching Research & Design*

Teaching *Research & Design* presents challenges for teachers, in terms of curriculum, pedagogy, assessment and organisation. Teachers need to support student learning while students engage in research and design activities, and they need to assess the progress their students are making, both in formative and summative ways. To date, limited research is available on how teachers implement integrated approaches to STEM education, partly because it is a relatively recent phenomenon (Stohlmann, Moore, & Roehrig, 2012). In a study on professional development needs for the implementation of integrated approaches to STEM education, Shernoff, Sinha, Bressler, and Ginsberg (2017) reported that teachers "did not know how to effectively integrate the STEM areas," and that "their lack of understanding of how to teach in integrated ways was strongly related to students' lack of understanding or lack of motivation to learn in different ways" (p. 8). Teachers also expressed that a shift in mindset was needed, away from pedagogies that aim to provide students with the 'correct answer' (Shernoff et al., 2017). Teachers of integrated STEM emphasised the importance of support and resources, such as time to prepare, implement and evaluate a project, or to work collaboratively with colleagues (Shernoff et al., 2017). Most teachers lack a preparation to teach integrated STEM, as most of them, especially primary teachers, are either prepared as generalists, or have a background in one particular discipline (Honey, Pearson, & Schweingruber, 2014). Requiring teachers to teach in content areas other than their own specialisation obviously creates challenges and knowledge gaps (Stinson, Herkness, Meyer, & Stallworth, 2009). Specifically, most teachers are not very familiar with design, how to teach and assess it, and how to connect research and design in student projects. The prominence of design, as a core practice in engineering, thus creates a specific challenge.

There is emerging empirical evidence for the effectiveness of the design process in facilitating the integration of concepts from multiple STEM areas (Estapa & Tank, 2017; Guzey, Moore, Harwell, & Moreno, 2016). Such findings relate to the abovementioned discussion about the balance between knowledge and skills in STEM education. For teachers, a focus on knowledge acquisition implies they need to be able to explain or clarify specific subject matter to their students, for instance, in relation to a STEM project they are conducting. Skill development, on the other hand, requires teachers to supervise (groups of) students and provide them with feedback on their progress. As the subject *Research & Design* aims at developing both content knowledge and skills, teachers need to be versatile, which includes being able to supervise students together with the STEM professionals from the industries and businesses who have initiated the R&D projects.

Previous research has shown that teachers generally hold positive attitudes towards contemporary teaching methods, like inquiry and design-based learning (Ara, Chunawala, & Natarajan, 2011). Recently, we, the authors of this chapter, conducted a yet unpublished study among a sample of 78 Dutch teachers of *Research & Design*, to explore their attitudes towards teaching research and design. These teachers had completed the required professional development program to be certified, and most of them (63) had been teaching R&D for 3 years or more. Their disciplinary backgrounds ranged from science and mathematics to arts and language. Findings indicated that teachers, overall, expressed positive attitudes. They enjoy supervising research and design projects and consider this a relevant activity. Interestingly, research was rated as slightly more relevant than design. Moreover, they rated their competence rather high (around 4 on a 5-point scale) and reported low values of anxiety (below 2). A positive attitude towards research and design is not automatically an indication that teachers are very knowledgeable. For instance, Allum, Sturgis, Tabourazi, and Brunton-Smith (2008) found in their meta-analysis only a very loose positive correlation between one's attitude towards science and one's understanding of science. Thus, the feeling of being competent to supervise research and design projects might not translate to teachers' actual knowledge of the research and design process or their enactment in practice. It is possible that R&D teachers do not deem it necessary to know a lot about the content of specific research and design activities, seeing themselves as facilitators of these processes in the first place. However, these findings are an indication that R&D teachers, from a variety of disciplinary backgrounds, have positive attitudes to teaching R&D. These attitudes might positively impact on their development towards becoming competent. Further research is underway to explore this.

4.2 *Students' Attitudes towards Research and Design*

Recently, Vossen et al. (2018) published a study on students' attitudes towards research and design, which was conducted among a sample of Technasium students (608 from Grade 8 and 314 from Grade 11), using a questionnaire based on the Dimensions of Attitude towards Science (DAS) instrument developed by Van Aalderen-Smeets and Walma van der Molen (2013). The authors concluded that students in secondary education had neutral to slightly positive attitudes towards doing research activities and somewhat more positive attitudes towards doing design activities. Across the board, they viewed design as less difficult and more enjoyable than research activities. Interestingly, students found research activities more relevant and important than design activities. Further research is needed to understand these differences; however, we think it is possible that students enjoy the open nature of design activities, which typically lead to unknown and new solutions. The positive attitude towards doing design activities reported by Vossen et al. (2018) is consistent with findings in other studies on students' attitudes towards engineering (Ara et al., 2011; Kőycű & de Vries, 2016) in which design was presented as a key activity within the engineering discipline. Other studies have provided empirical evidence for an association of design activities with positive attitudes towards STEM careers and skills like problem solving, creativity, communication and teamwork (e.g., Guzey et al., 2016; Moore, Glancy, Tank, Kersten, Smith, & Stohlmann, 2014).

Vossen et al. (2018) also compared the attitudes of Technasium students, who are taking the subject R&D, with students from other schools that did not offer R&D as a subject but provide research (and sometimes design) activities as part of the 'traditional' science subjects. The authors reported that R&D students had significantly more positive attitudes towards doing design activities than non-R&D students. Also, compared to non-R&D students, R&D students found doing research activities significantly more relevant, and experienced less anxiety towards doing research tasks. R&D students scored significantly higher on aspirations to pursue a design related study or career, whereas in the non-R&D group, students scored significantly higher on interest in a research related future occupation. This difference could be explained by the fact that only R&D students have had extensive experiences with performing design activities and apparently find these more enjoyable than doing research. The results of this study suggest that a project-based subject like R&D can enhance students' attitudes towards the core STEM practices, that is, research and design. However, it is known from other research, that positive attitudes and enjoyment do not necessarily translate into post-16 STEM participation in studies, or careers (Archer, Osborne, DeWitt, Dillon, Wong, & Willis, 2013).

5 Implementing STEM Education: Collaboration Is Key

In the Technasium example discussed above, schools are committed to a framework that requires investing in a studio, providing curriculum time for R&D, participating in a network, and making sure that teachers of R&D are certified. Within this framework, however, each school creates their own version of the Technasium concept. Specific projects depend on local industries, and the interests and capabilities of students and teachers. Consequently, some schools have chosen to focus on engineering, or on specific technologies. Regardless of the choices or priorities of Technasium schools, there is always a need for teachers of various backgrounds and specialisations to design curriculum materials and tasks together, and to negotiate the use of specific pedagogies and assessments, for instance, to obtain some consensus on how to teach and assess certain skills. In addition, R&D teachers meet with colleagues from other Technasium schools in their network to learn from each other's experiences. Also, R&D teachers meet with STEM professionals from local industries to coordinate the supervision of student projects. In other words, collaboration is paramount in any variation of Technasium.

We think that the Technasium example contains some interesting features. First, the school-industry partnerships provide students with opportunities to experience real world STEM problems or issues. Second, the focus on research and design projects implies an interdisciplinary approach that requires students to work together, but also that teachers need to connect with colleagues and STEM professionals who have complementary expertise. Third, the Technasium concept is not a straightjacket but a framework that allows schools to create their own models or versions. We argue that providing students with real world STEM experiences is a key to effective STEM education in general. Such experiences will assist students in making informed decisions about embarking on STEM related tertiary studies or careers (Education Council, 2018). To provide such experiences, we propose that STEM education in general will benefit from linkages between schools and external parties. Many teachers have never worked outside a school. Connecting with local STEM industries will help them to understand the reality of STEM professionals and broaden and inspire their view of STEM education. Connecting with local industries implies that, at the school level, STEM education may look quite different in different places. It can be more or less interdisciplinary or integrated, depending on the priorities that schools and teachers set for their population of students, and on the partnerships and the resources that are available to them. School leadership is required to support and foster collaboration between teachers and STEM professionals. Since in most schools around the world, secondary schools in particular, collaboration across disciplines or school subjects is not very common, coaching for multidisciplinary teams of teachers may be necessary.

School-industry partnerships can have significant benefits for schools, teachers, and most importantly, students. (Education Council, 2018). Reviewing such partnerships, Hobbs, Jakab, Millar, Prain, Redman, Speldewinde, Tytler, and van Driel (2017) concluded that "exposure to the human aspects of practice in STEM industry … through partnerships can engage and motivate both male and female students to consider studying and having a career in STEM. This is particularly pertinent to engaging girls with STEM through contact with strong female STEM practitioner role models" (Hobbs et al., 2017,

p. 23). However, this study also warned that schools and industries are different worlds, each with their own purposes and cultures, and that short-term initiatives often result in initiatives having no systemic impact beyond the project period. The Education Council (2018) concluded that "to be effective, partnerships have to address issues related to commitment, personnel, bridging different contexts and language, and recognise the primary role that principals and teachers have in defining the pedagogy and teaching in their schools" (p. 9). We thus recommend the establishment of transformative partnerships, on-going relationships between schools and industries, well embedded in the school context, that offer mentoring for students and sharing of expertise for teachers.

6 Conclusion

In this chapter, STEM education was used to refer to curricula or subjects that were purposefully developed to connect the STEM disciplines. In particular, we discussed an approach to STEM education that focuses on connecting research and design as core practices across the STEM disciplines. The subject *Research & Design* in the Dutch Technasium schools aims to provide students with real world STEM experiences through school-industry partnerships. Research on attitudes of Technasium students and teachers indicated that students perceive design as less difficult and more enjoyable than research activities, however, the latter were considered more relevant and important than design activities. Teachers reported to enjoy and feel competent to supervise research and design projects, however, they rated research as slightly more relevant than design. Collaborations between teachers, within and across schools, and with STEM professionals working in local industries are an essential element in the implementation of this innovation.

On the basis of this exploration, we conclude that STEM education can be effective in terms of engaging students and increasing their awareness of STEM practices, if it provides:

- Partnerships that link schools (primary, secondary, vocational) with STEM industries and businesses in sustained ways;
- Opportunities for teachers to collaborate with and learn from peers and STEM professionals and;
- Provisions for research to explore the impact of STEM education. Research should not be limited to counting numbers of students who take up STEM subjects, but needs to have a broader scope, including changes in student attitudes and teacher practices.

Rather than prescribing a national STEM curriculum, STEM education can benefit from a framework that is set at a national or state level, and which leaves room for schools and teachers to develop and implement a version of STEM education that fits their local context, student population and resources. Changes at the levels of students, parents, teachers, schools and industries will not happen overnight, and will require sustained efforts. Therefore, a long-term research program is needed to investigate the impact of STEM education at these various levels, focusing on changes in practice and attitudes.

References

Allum, N., Sturgis, P., Tabourazi, D., & Brunton-Smith, I. (2008). Science knowledge and attitudes across cultures: A meta-analysis. *Public Understanding of Science, 17*(1), 35–54.

Ara, F., Chunawala, S., & Natarajan, C. (2011). A study investigating Indian middle school students' ideas of design and designers. *Design and Technology Education: An International Journal, 16*(3), 62–73.

Archer, L., Osborne, J., DeWitt, J., Dillon, J., Wong, B., & Willis, B. (2013). *ASPIRES report: Young people's science and career aspirations, age 10–14*. London: King's College.

Bruning, L., & Michels, B. (2014). *Handreiking schoolexamen Onderzoek & ontwerpen in de tweede fase* [Instruction manual for school exams research & design in upper secondary education]. Retrieved from http://www.slo.nl/organisatie/recentepublicaties/handreikingonderzoek/

Bybee, R. W. (2010). Advancing STEM education: A 2020 vision. *Technology and Engineering Teacher, 70*(1), 30–35.

Creswell, J. W. (2008). *Educational research: Planning, conducting, and evaluating quantitative and qualitative research* (3rd ed.). Upper Saddle River, NJ: Pearson.

de Vries, M. J. (2005). *Teaching about technology: An introduction to the philosophy of technology for non-philosophers*. Dordrecht: Springer.

Education Council. (2018). *Optimising STEM industry-school partnerships: Inspiring Australia's next generation, final report*. Retrieved from http://www.educationcouncil.edu.au/EC-Reports-and-Publications.aspx

Estapa, A. T., & Tank, K. M. (2017). Supporting integrated STEM in the elementary classroom: A professional development approach centered on an engineering design challenge. *International Journal of STEM Education, 4*(1), 6. doi:10.1186/s40594-017-0058-3

Fallman, D. (2003, April). *Design-oriented human-computer interaction*. Proceedings of the ACM CHI 2003 Human Factors in Computing Systems Conference, Fort Lauderdale, FL.

Guzey, S. S., Moore, T. J., Harwell, M., & Moreno, M. (2016). STEM integration in middle school life science: Student learning and attitudes. *Journal of Science Education and Technology, 25*(4), 550–560.

Hathcock, S. J., Dickerson, D. L., Eckhoff, A., & Katsioloudis, P. (2015). Scaffolding for creative product possibilities in a design-based STEM activity. *Research in Science Education, 45*(5), 727–748.

Hobbs, L. (2013). Teaching 'out-of-field' as a boundary-crossing event: Factors shaping teacher identity. *International Journal of Science and Mathematics Education, 11*(2), 271–297.

Hobbs, L., Jakab, C., Millar, V., Prain, V., Redman, C., Speldewinde, C., Tytler, R., & van Driel, J. (2017). *Girls' future – Our future: The Invergowrie Foundation STEM report.* Melbourne: Invergowrie Foundation. Retrieved from http://www.invergowrie.org.au/girls-future-our-future-the-invergowrie-foundation-stem-report/

Honey, M., Pearson, G., & Schweingruber, A. (2014). *STEM integration in K-12 education: Status, prospects, and an agenda for research.* Washington, DC: National Academies Press.

Ingersoll, R. M. (1998). The problem of out-of-field teaching. *Phi Delta Kappan, 79*(10), 773–776.

Johnson, C. C. (2013). Conceptualizing integrated STEM education. *School Science and Mathematics, 113*(8), 367–368.

Kaspura, A. (2017). *Engineers make things happen: The need for an engineering pipeline strategy.* Barton: Institution of Engineers Australia. Retrieved from https://www.engineersaustralia.org.au/Pitch-Better-Nation/Make-Things-Happen

Kennedy, J. P., Lyons, T., & Quinn, F. (2014). The continuing decline of science and mathematics enrolments in Australian high schools. *Teaching Science, 60*(2), 34–46.

Kolodner, J. L., Gray, J., & Fasse, B. B. (2003). Promoting transfer through case-based reasoning: Rituals and practices in learning by design classrooms. *Cognitive Science Quarterly, 3*(2), 119–170.

Kőycű, Ü., & de Vries, M. J. (2016). What preconceptions and attitudes about engineering are prevalent amongst upper secondary school pupils? An international study. *International Journal of Technology and Design Education, 26*(2), 243–258.

Marginson, S., Tytler, R., Freeman, B., & Roberts, K. (2013). *STEM: Country comparisons.* Melbourne: The Australian Council of Learned Academies. Retrieved from http://acola.org.au/wp/PDF/SAF02Consultants/SAF02_STEM_%20FINAL.pdf

Mehalik, M. M., Doppelt, Y., & Schuun, C. D. (2008). Middle-school science through design-based learning versus scripted inquiry: Better overall science concept learning and equity gap reduction. *Journal of Engineering Education, 97*(1), 71–85.

Moore, T. J., Glancy, A. W., Tank, K. M., Kersten, J. A., Smith, K. A., & Stohlmann, M. S. (2014). A framework for quality K-12 engineering education: Research and development. *Journal of Pre-College Engineering Education Research (J-PEER), 4*(1), 2. doi:10.7771/2157-9288.1069

National Research Council (NRC). (2012). *A framework for K-12 science education: Practices, cross-cutting concepts, and core ideas.* Washington, DC: National Academies Press.

NGSS Lead States. (2013). *Next generation science standards: For states, by states.* Washington, DC: National Academies Press.

Shernoff, D. J., Sinha, S., Bressler, D. M., & Ginsburg, L. (2017). Assessing teacher education and professional development needs for the implementation of integrated approaches to STEM education. *International Journal of STEM Education, 4*(1), 13. doi:10.1186/s40594-017-0068-1

STAR portal. (2017). Retrieved from https://starportal.edu.au/

Stinson, K., Harkness, S. S., Meyer, H., & Stallworth, J. (2009). Mathematics and science integration: Models and characterizations. *School Science and Mathematics, 109*(3), 153–161.

Stohlmann, M., Moore, T. J., & Roehrig, G. H. (2012). Considerations for teaching integrated STEM education. *Journal of Pre-College Engineering Education Research (J-PEER), 2*(1), 4. doi:10.5703/1288284314653

Van Aalderen-Smeets, S., & Walma van der Molen, J. (2013). Measuring primary teachers' attitudes toward teaching science: Development of the Dimensions of Attitude toward Science (DAS) instrument. *International Journal of Science Education, 35*(4), 577–600.

Vossen, T. E., Henze, I., Rippe, R. C. A., van Driel, J. H., & de Vries, M. J. (2018). Attitudes of secondary school students towards doing research and design activities. *International Journal of Science Education, 40*(13), 1629–1652. doi:10.1080/09500693.2018.1494395

Williams, J., Eames, C., Hume, A., & Lockley, J. (2012). Promoting pedagogical content knowledge development for early career secondary teachers in science and technology using content representations. *Research in Science & Technological Education, 30*(3), 327–343.

Willison, J., & O'Regan, K. (2008). *The researcher skill development framework.* Retrieved from http://www.adelaide.edu.au/rsd2/framework/rsd7/

CHAPTER 3

Reading STEM as Discourse

Kathy Jordan

Abstract

A STEM discourse is emerging in Australian national and state school education policies, as governments seek to develop a vision and road map for the future. This official discourse argues in support of a national STEM enterprise, so that Australia's economic growth and way of life will be maintained. Within this discourse, schools are framed as important to increasing the participation and performance of students in STEM, and to building a highly-skilled STEM workforce. This chapter analyses several recent national and state school education policies to gain greater understanding of this positioning.

Keywords

Discourse – policy – text – power

1 Introduction

In 2013 the Chief Scientist released the position paper, *Science, Technology, Engineering and Mathematics in the national interest: A strategic approach,* calling for a whole of government approach to investment in STEM because of perceived "gathering threats" and "challenges" to the country's "continued prosperity." The role of the Chief Scientist, is to provide high level advice to the prime minister and other ministers relating to science, technology and innovation. Following considerable broad support for this paper, the Chief Scientist released another paper in 2014, *Science, Technology, Engineering and Mathematics: Australia's future,* as a recommendation to government. This later paper has been highly influential, leading to the development of the *National STEM School Education Strategy* (2015) as well as the Victorian state policy, *STEM in the Education State* (2016).

Policies are a means by which governments and organisations "articulate preferred visions of the future, and are used to make decisions in the present that are consistent with these visions" (Moyle, 2005, p. 3). Policies such as those

 | DOI:10.1163/9789004391413_004

above relating to STEM in school education, present a formal sanctioned view and set the direction or action of government to achieve specific objectives. They have a theoretical and a logistical duality: that is, they both articulate a vision for a preferred future and provide for its operational application (Lloyd, 2008). Policies however are socially constructed, and present particular ways of thinking, and acting.

Policy can be a term that is taken for granted and can be used to describe very different things (Ball, 1993). To Ball, policy has a dual conceptualisation, as text as well as discourse. Perceiving policy as text, Ball draws on the influence of literary theory to argue that policies are representations that are always changing and are never complete, as a researcher is always dealing with one policy in relation to another and other responses to that policy (Fimyar, 2014). He argues that they are the products of multiple authors, agendas and decision-making and as such are what he terms as 'cannibalized products' (p. 16). Perceiving policy as discourse Ball draws on the influence of Foucault to argue that policy is also a power relation in which power is exercised through it. Thus, to Ball policy as discourse emphasises the value-laden nature of policy. As commented by Ball (1993), policies,

> Exercise power through a production of 'truth' and 'knowledge,' as discourses … Discourses are about what can be said, and thought, but also about who can speak, when, where and with what authority. Discourses embody the meaning and use of propositions and words. Thus, certain possibilities for thought are constructed. Words are ordered and combined in particular ways and other combinations are displaced or excluded. (p. 14)

It is important for us to read policy carefully, to understand both the vision being proposed and the roadmap presented for us to follow. Powerful groups including policy makers and politicians can control the knowledge presented and also condition readers to specific representations and accounts of policies, issues and debates. Policy makers can and do constrain the way we think about issues, ideas and concepts through the language which they use to frame them (Fimyar, 2014). The use of the language of business, marketing and finance is often used, so too a focus on outcomes, knowledge and skills. Policies can also become self-fulfilling. Whether intended or not, policies shape understandings and expectations (Selwyn, 2011) and in turn impact the stories or narratives that we construct and act upon. Critically reading policies draws attention to the ideologies of what is being said and indeed what is not being said.

This chapter focuses on analysing recent national and state policies relating to STEM in school education, as a means of furthering our understanding of the conceptualisation of STEM in this specific context. It takes the view that all conceptualisations or all stories that we construct need to be scrutinised. And here perhaps a cautionary tale is perhaps useful. For well over forty years, the perceived benefits of educational technologies have become normalised into everyday thinking, so that there is now an orthodoxy around their use (Selwyn, 2011). Technologies are now accepted and taken for granted as a good thing, and it would take a very brave teacher to suggest that they were not. However, this accepted discourse is very different from the discourse that first emerged in the 1990's, thus demonstrating how a discourse can subsume competing discourses so that only one discourse remains. As commented by Selwyn (2011),

> Whereas a distinct resistance to information technologies in some segments of the educational community may have existed during the 1980s and 1990s, most academics and practitioners have now moved beyond a disinterested acceptance of digital technology to a deep-rooted and widely held belief in the inherent befits of technology for education. As such, digital technology use has now achieved the status of an integral and institutionalised part of the fabric of contemporary education – something that barely requires thinking about, or even, acknowledging. (p. 20)

Over the last forty years however many of the claims and promises for educational technologies have not been realised. While the rhetoric espoused technologies as revolutionising teaching and learning, much of the actual use in schools is to make teachers day to day operations quicker, faster and easier. As such, proponents of a STEM agenda need to be mindful that official dominant discourses need to be tempered with the messy realities of use in schools.

2 Analysing Discourse

The term discourse is used in our everyday language and used interchangeably to refer to discussion or dialogue. A discourse analysis therefore refers to an analysis of discourses, the patterns and rules and purposes in how language is used (Hewitt, 2009). The overall aim of discourse analysis then is to understand how a discourse constructs a certain reality. All literacy practices are socially constructed in and through language (Gee, 2011) and language shapes and determines ideas and actions. Language is also politicized.

The objective of analysing discourse is to uncover the assumptions that are conveyed in the language we use. Witness for example how the words "stop the boats" have particular connotations for Australians following 2001, when then Prime Minister John Howard, introduced the Pacific Solution, and the use of offshore processes and temporary via protection to deter asylum seekers. Or consider how the words "fake news" have particular connotations, given its use by United States president Donald Trump to dismiss and to discredit contrary views to those he wishes to promote.

There is no one theoretical framework or methodology to use when analysing discourse. Discourse analysis is used in many disciplines including linguistics, education and social sciences each of which has its own definitions, practices and methodologies. Linguistic traditions define discourse based on units of written and spoken communication under study and therefore focus on the content of texts and conversations. Other social science traditions, define discourse as being dependent on social practice and how these form rules which work together to construct discourse (Gee, 2005; Van Dijk, 2015). This view of discourse, in arguing that discourses are formed and shaped, also argues that discourses can be reshaped. Different disciplines have different modes of discourse analysis with some transferring across disciplines leading to a multiplicity of approaches.

According to Van Dijk (2015) discourse analysis is premised on studying "the way social-power abuse and inequality are enacted, reproduced, legitimated, and resisted by text and talk in the social and political context" (p. 466). He sees the goal of discourse analysis is to expose and ultimately transform social inequality by making evident the connections between language and power. By making the discourse explicit, the intent is to support the oppressed and encourage resistance and transformation of lives (Lankshear, Snyder, & Green, 2000). Engaging in a discourse analysis then can be seen as a means to question the voice of those in power, to reveal motives and agendas that can serve self-interest and to enable the voices of the marginal to be legitimised (Van Dijk, 2015).

> Discourses are not merely talk; they are performative and produce particular versions of social reality to the exclusion of other possibilities, thereby substantially shaping socio-economic, institutional, and cultural conditions and processes. (Greckhamer & Cilesiz, 2014, p. 424)

There are many different approaches to discourse analysis, with some researchers looking at the content of the language, the themes and issues and some looking at the structure of language and how it functions to make meaning in specific contexts. Van Dijk (2015) suggests that when reading a text, we need to read with a critical eye, to raise questions about it, imagine how it could have been constructed and how it could have been constructed differently, check the perspective being presented, its angle or slant, and how it is framed. It is also about looking at the images included, the placement of photographs, the use of headings, key words, and the selected voice and the tone. He draws on his own research around the discourse of immigration to suggest some common discourse structures that can be used to position the text and reading of that text. These are shown in Table 3.1.

Gee (2005) takes the view that discourse analysis needs to be critical, as language itself is political. According to Gee (2011) to do discourse analysis requires us to see "what is old and taken for granted as if it were brand new. We need to see all the assumptions and information speakers leave unsaid and assume listeners know and will add in to make the communications clear" (p. 8). Gee (2005, 2011) identified interrelated aspects of reality that we construct through discourse, namely: significance, activities, identities, relationships, politics, connections, sign systems and knowledge. These building blocks with corresponding discourse analysis questions are shown in Table 3.2.

TABLE 3.1 Common discourse structures that can be used to position the text and reading of that text

Headlines and leads	Can express main topics as defined by the writer. A demonstration can be positioned as a violation of the social order or as a democratic right of demonstrators.
Implications and presuppositions	Can assert facts that may or may not be true. The acts of demonstrators can be asserted as 'violent.'
Metaphors	Can make the abstract more concrete. The metaphor 'wave' of immigrants can create fear of drowning in massive numbers of immigrants.
Lexicon expression	Can influence opinions. Migrants can be labelled as 'illegal' or 'undocumented.'
Passive sentence structures and nominalizations	Can hide or down play. Discrimination can be mentioned in a non-explicit way.

TABLE 3.2 Building blocks with corresponding discourse analysis questions

Building tasks	Discourse analysis questions
Significance	How is this piece of language being used to make certain things significant or not and in what ways?
Activities	What practice (activity) or practices (activities) is this piece of language being used to enact (i.e., get others to recognize as going on)?
Identities	What identity or identities is this piece of language being used to enact (i.e., get others to recognize as operative)? What identity or identities is this piece of language attributing to others and how does this help the speaker or writer enact his or her own identity?
Relationships	What sort of relationship or relationships is this piece of language seeking to enact with others (present or not)?
Politics	What perspective on social goods is this piece of language communicating (i.e., what is being communicated as to what is taken to be "normal," "right," "good," "correct," "proper," "appropriate," "valuable," "the ways things are," "the way things ought to be," "high status or low status," "like me or not like me," and so forth)?
Connections	How does this piece of language connect or disconnect things; how does it make one thing relevant or irrelevant to another?
Sign systems and knowledge	How does this piece of language privilege or disprivilege specific sign systems (e.g., Spanish vs. English, technical language vs. everyday language, words vs. images, words vs. equations, etc.) or different ways of knowing and believing or claims to knowledge and belief (e.g., science vs. the Humanities, science vs. "common sense," biology vs. "creation science")?

3 Approach

This chapter focuses on a discourse analysis of the policy documents acknowledged in this introduction to this chapter, as a means of furthering our understanding of the conceptualisation of STEM as part of the vision and roadmap being articulated for school education at the national and Victorian state level. Any discourse analysis is not based on all the physical features pertaining to a text or conversation. Rather, it requires the researcher to make judgements of relevance around what should be included and what should not. Any analysis of a text is therefore analysed along a continuum from very detailed to narrow.

Given that this chapter focuses on several national and state policies, the decision was made to make this analysis a narrow one. It broadly draws on Gee's building blocks (2005, 2011) and his advice around picking key words and phrases in the data and looking for situated meaning and the discourses that seemed to be relevant as well as the linguistic details that appear to be appropriate.

I begin by introducing each of the policies and their main purpose. I then turn to consider the key themes around STEM that are represented in these policies and use them as a framework to guide subsequent discussion. The focus of analysis is on the representation of STEM – such as the language used, the tone, the use of images, and headings – discussion of the actual strategies and recommendations being proposed for STEM are outside the scope of this chapter unless they are relevant to this representation.

4 Introducing the Policies

The paper, "Science, Technology, Engineering and Mathematics in the national interest: A strategic approach" was released in 2013 (Office of the Chief Scientist, 2013). The intent of this policy, as clearly stated in its title, is to propose a national 'position' for STEM. Hereafter this paper is referred to as the position paper.

Following favourable response to this position paper, the Chief Scientist produced "Science, Technology, Engineering and Mathematics: Australia's future" (Office of the Chief Scientist, 2014), hereafter referred to as the recommendations paper. As stated in the introductory sections of the paper, entitled 'An agenda for change,' its intent is to build on the previous position paper and put forward 'a strategic approach to STEM as a recommendation to government' (2014, p. 1). These recommendations involved four areas of focus, referred to as four 'means' (2014, p. 6): Australian competitiveness, Education and training, Research and International engagement.

The third policy paper "National STEM School Education strategy" (Education Council, 2015) was developed in response to this recommendation paper (Office of the Chief Scientist, 2014) and was endorsed by the Education Council in 2015. The Education Council was formed in 2013 by the Council of Australian Governments (COAG) to enable strategic policy on school education, early childhood and higher education to be coordinated at the national level, and to provide a mechanism to share information and use resources collaboratively. The intent of this policy, referred to as the national strategy (Education Council, 2015) is made clear by its title, that is, to provide a comprehensive national plan for STEM Education.

The final policy is the Victorian state strategy, *STEM in the Education* State, that was published by the Department of Education and Training in 2016 (Department of Education and Training, 2016) and is referred to as the state strategy. The intent of this paper, again as iterated in the title, is to articulate how STEM is being enacted in Victoria.

It is interesting that while the focus of each of these policies is on STEM, there is not much attention given to defining STEM and is often presented as self-evident. Both the position paper (2013) and the recommendations paper (2014), refer to the "Scope of STEM" in an appendix. In the state strategy (2016), STEM refers to disciplines and as generic skills associated with them including creativity and problem solving. The national strategy (2015) does define STEM,

> To refer collectively to the teaching of the disciplines within its umbrella – science, technology, engineering and mathematics – and also to a cross-disciplinary approach to teaching that increases student interest in STEM-related fields and improves students' problem solving skills. (p. 5)

5 Representations of STEM

The ensuring discussion turns attention to the ways that STEM is represented in the selected national and state policy documents. This discussion is organised by the broad themes that emerge from this analysis.

5.1 *Theme 1: STEM as National Enterprise*

An enterprise refers to a project or undertaking, especially a bold one. It is synonymous with a venture or operation and involves initiative and resourcefulness. The term is used explicitly in the position paper (2013) to refer to both the proposition, that is, a national agenda for STEM and the outcome of this proposition, a prosperous future, and a valued undertaking by the Australian community.

The national strategy (2016) is focused on the achievement of two goals; to "Ensure all students finish school with strong foundational knowledge in STEM and related skills" and to "Ensure that students are inspired to take on more challenging STEM subjects." To achieve these goals, the plan includes five areas for national action through school education. These are:

1. Increasing student STEM ability, engagement, participation and aspiration,
2. Increasing teacher capacity and STEM teaching quality,
3. Supporting STEM education opportunities within school systems,

4. Facilitating effective partnerships with tertiary education providers, business and industry and
5. Building a strong evidence base.

A common format is used for all 5 action areas. This includes the use of numbering (from 1 to 5 to indicate the definitive areas of action), bold text (to draw reader attention), direct language, such as "quality teaching is the key," and reference to proof such as "evidence shows" and "research shows" (to create confidence and assurance). Each of the 5 action areas include the same two sub-sections, National collaborative actions and Jurisdictional priority actions. These are presented in a different colored font, with text presented in a series of dot points. However key words associated with each sub-section are quite different. Words such as "explore," "increase," "establish," "extend," and "develop," are used in relation to the National collaborative actions. A more collaborative language, through words such as "supporting," "encouraging" and "continuing," is used in relation to the Jurisdictional priority actions.

Linked with this notion of enterprise is calls for a national reform agenda or national strategic approach to STEM. The three nationally focused policies, support this agenda, as a means of drawing together the programs already undertaken. For example, the position paper (2013) comments that, "It is not that we lack programs in Australia. There have been many programs, large and small, built over decades. What we do lack is a national approach to STEM" (p. 10). The recommendations paper (2014) in arguing similarly adds that "Australia is now the only country in the OECD not to have a current national strategy that bears on science and/or technology and/or innovation" (p. 10). Furthermore, it suggests that there has been a lack of 'coordination, misdirected effort, instability and duplication' leading to a "patchwork of programmes relevant to STEM" (Office of the Chief Scientist, 2014, p. 10). The purpose of the national strategy (Education Council, 2015) is to solve this hotchpotch approach through collaboration,

> There is significant activity underway across the country – within schools, school systems, universities and business – to improve STEM education. The national STEM school education strategy seeks to build on this activity and provides a framework for collaborative effort. (p. 6)

5.2 *Theme 2: STEM as Sustaining Economic Growth*

STEM is commonly perceived as vital to ensuring the health and vitality of Australia's continued economic prosperity, and as a means to achieving this end. This is particularly evident in the recommendations paper (2014).

It comments for example how "the global economy is changing" and how STEM is "the almost universal preoccupation now shaping economic plans" (2014, p. 5) in other countries, and that investment in STEM is needed "to build a stronger Australia with a competitive economy" and for performing a "pivotal role in securing a stronger Australia" (2014, p. 6). Underlying is a sense of urgency and fear of not keeping up with other countries as illustrated in the following comment, "Australians must decide whether we will be in the forefront of these changes or be left behind. We have a choice" (2014, p. 5).

In the position paper (Office of the Chief Scientist, 2013) this advocated position is presented as an absolute one; with the word "must" used to position the reader of such. Fear of consequences, such as to the economy, and to our way of life are also used as shown in the following comment. "The reality is that we can't relax. We can't be complacent. There can be no sense of entitlement. We must understand that we will get the future we earn" (Office of the Chief Scientist, 2013, p. 3). School education is positioned around providing future STEM practitioners for a STEM literate society, to be achieved by encouraging enrolment and recruitment of STEM teachers and increasing time for science in primary schools.

The recommendation paper (Office of the Chief Scientist, 2014) is underpinned by a sense of urgency and fear of consequences if national action around STEM is not taken. For example, larger sized bold text is used for emphasis, and matter of fact language, that does not invite an alternate view, is used to frame each of the initial sections of this report. The language used in the titles of these earlier sections, including "Facing up to the task," "The ends and the means," "Too important to leave to chance," and "Too long left to chance" aim to position the reader to believe that action must be taken. These opening sections also use language of "choice" including the following phrases "Australians must decide," and "we have a choice" (Office of the Chief Scientist, 2014, p. 5), to position readers to perceive themselves as empowered to make decisions.

The state strategy (Department of Education and Training, 2016) also picks up on this theme and comments similarly on the changing economy and the need for STEM skills. This is illustrated in the following comment that "significant changes in Victoria's economy mean there is a greater need for STEM capabilities than ever before' (2016, p. 1). However, this policy also explicitly connects perceived economic benefits to education, commenting thus, "Ensuring our learners are well-equipped with STEM skills and knowledge – and have the confidence and enthusiasm to use them – will help secure Victoria's future as a competitive, innovative and vibrant economy" (2016, p. 1). Thus, a simple cause and effect relationship between STEM and the economy is framed.

The State strategy (Department of Education and Training, 2016) incorporates three sections. The first section, "STEM in the Education State" builds a case for investment in STEM education and uses matter of fact language to do so. It is accompanied by further text describing the "ambitious targets" (Department of Education and Training, 2016, p. 4), the "commitment" (Department of Education and Training, 2016, p. 4), and the "aims" (Department of Education and Training, 2016, p. 5) of the government agenda. Graphs, or other visual images of data, is used to build the argument that Victoria is currently underperforming in mathematics (by year level, in comparison to selected states and countries, and by socio-economic status), and that students have low perceptions of science (by year level and gender) as well as low participation rates in STEM at tertiary level. The second section, entitled "We are taking action to improve STEM education and skills," functions as a type of report card, a means for reporting on successful government initiatives and projects. Here the use of the word "we" is quite prominent, such as "we are creating," "we are investing" and "we are helping" (Department of Education and Training, 2016, p. 10), "we are supporting" (ibid., p. 11), "we are building" and "we are transforming" (ibid., 2016, p. 13). In the third section, entitled, "What's next for STEM in the Education State?" the focus is on reporting additional initiatives, where the language is future orientated through words such as "will" to create a sense of confident assurance that the government plan will continue to be successful. The tone here is upbeat, confident and under control.

5.3 *Theme 3: STEM as Maintaining Prosperity*

STEM is portrayed as the means to ensure the "continued prosperity of Australia on all fronts – socially, culturally and economically – for all our citizens and for our place in the world" (Office of the Chief Scientist, 2013, p. 7). The position paper (Office of the Chief Scientist, 2013) does this by using a cause and effect argument. Firstly, it describes Australia's current economic prosperity, as shown in this comment, "by many measures, the Australian community is one of the most successful in the world. With extensive natural resources, served by fair, stable and respected institutions, and characterised by ingenuity and hard work, many Australians are adequately prosperous" (2013, p. 5).

Secondly, it suggests that this prosperity is subject to "gathering threats and challenges" (Office of the Chief Scientist, 2013, p. 5), which it then lists. These being: living in a changing environment, promoting population health and wellbeing, managing our food and water assets, securing Australia's place in a changing world, and lifting productivity and economic growth. Thirdly, it suggests that STEM is the means to address these threats and challenges (Office of the Chief Scientist, 2013, p. 5), iterating that this is what STEM has always done as shown in this comment, "STEM has and will continue to provide for everyone – to make available the new knowledge and technologies that are needed to address these challenges" (Office of the Chief Scientist, p. 5). And this, that "Without this wellspring of new knowledge, there are hard limits to the potential for further improvement of peoples' lives" (Office of the Chief Scientist, p. 5). Thus, STEM is perceived as an economic and social leveller.

5.4 *Theme 4: STEM as Not Being Left Behind*

Related to this notion of sustaining our economic future and prosperity is a sense of the need for urgent national action, for fear that inaction will lead to Australia being left behind. The position paper (Office of the Chief Scientist, 2013) writes about this in this way, that

> Many countries are relying on their STEM enterprise and its quality to build knowledge-based communities and economies. Australia must do the same. There is an urgent need to act if we are not to be left behind. (p. 6)

And this, "around the world there is a sense of urgency – a need to improve a nation's capacity and a commitment not to take the future for granted" (Office of the Chief Scientist, 2013, p. 9). It also suggests that taking action is a sign of good or effective government as shown in the following comments, "Not being left behind as a nation and designing our future with STEM as a critical element ought to be the goals of governments" (Office of the Chief Scientist, 2013, p. 9) and this, "similar sentiments are expressed in developed and developing economies. And governments act" (Office of the Chief Scientist, p. 9), as well as this, "anxiety about being left behind creates a sense of urgency" (Office of the Chief Scientist, p. 9), and finally this, that

> Most nations are closely focused on advancing STEM, and some have evolved dynamic, potent and productive strategies. In world terms Australia is positioned not far between the top group, but lacks the national urgency found in the United States, East Asia and much of Western Europe, and runs the risk of being left behind. (p. 10)

The recommendations paper (Office of the Chief Scientist, 2014) similarly comments that, "It is time to do what so many other countries have already done: take a long-term strategic view of STEM's pivotal role in securing a stronger Australia" (p. 6).

5.5 *Theme 5: STEM as Securing a Workforce*

Securing a workforce with the perceived skill set required for a future workplace is also featured in these policies, with school education seen as the supplier. Both the position paper (Office of the Chief Scientist, 2013) and recommendations paper (Office of the Chief Scientist, 2014) make this link, with the national strategy (Education Council, 2015) also commenting on the growth rate in STEM-related jobs and predicted future employment in these areas. Yet it is the state strategy (Department of Education and Training, 2016) that focuses attention on work. It comments, for example that, "For Victoria, the key to a prosperous future lies in a highly-skilled workforce, including strong capability in STEM" (Department of Education and Training, 2016, p. 3). And this, "our employers are increasingly looking for workers who are creative problem solvers, innovative and creative thinkers, and able to use new technologies" (p. 1). Furthermore, it suggests that STEM skills will be needed for some 400, 000 jobs for Victorians by 2025 (Department of Education and Training, 2016, p. 1) and therefore has made a commitment to ensuring that students gain these necessary skills in schools, so as to pursue further study and careers. According to the state strategy (Department of Education and Training, 2016) employers are also seeking a workforce who think creatively and critically, and work collaboratively to solve problems.

5.6 *Theme 6: STEM as Declining*

The four policy documents each give attention to declining participation rates and performance levels in STEM. For example, the position paper (Office of the Chief Scientist, 2013) in advocating for a national and whole-of-government approach to STEM, argues that there is a lack of time in science in primary schools, that there is declining interest in STEM in secondary schools, as well as in tertiary study and STEM skill shortages in the workforce. It advocates for school education to offer incentives for enrolment and recruitment strategies, and increasing time spent on teaching science, as mentioned earlier.

The recommendations paper (Office of the Chief Scientist, 2014) focuses attention more on using a matter-of-fact tone to describe declining performance in mathematics at the international level, the high proportion of teachers who teach out of field, and the decline in science performance and rates of participation in science. It advocates for inspirational teaching and inspired learning.

The national strategy (Education Council, 2015) acknowledges that there are concerns about participation and performance and cites the recommendations paper (Office of the Chief Scientist, 2014) but does not refer further by giving an in-depth description of this decline. Rather it focuses on how much work has been done to raise the participation and performance levels in STEM in school education. It takes a conciliatory approach, that aims to "build on reforms and activities already underway" and that "reversing the trends" (Education Council, 2015, p. 3) in STEM performance and participation are complex and that many factors come into play including community and parent views, industry views and university admission policies. It puts forward two goals focused on increasing participation and performance, as stated earlier in this chapter.

6 Conclusion

The STEM discourse is future orientated and presents a positive vision for this future, one characterised by hope and promise in which STEM is positioned as the means for Australia to maintain its economic advantage and for Australians to maintain their prosperous standard of living. This vision of the future is dependent on school education. School education is seen as playing a vital role in developing students' knowledge and skills and as necessary to pursue further education in this field and to secure jobs in the future workplace; one that is characterised by STEM. This vision of the future is presented as relatively simple and straight forward. For example, increased participation and performance is presented as readily achievable by increasing the amount of time given over to the teaching of science in primary schools and through 'inspirational teaching' that will motivate students. While lifting participation and performance is seen as involving the community, government, parents etc, it is schools and teachers who are perceived as having a key role. Thus, it can be relatively easy then when participation and performance does not increase for schools and teachers to be easily blamed.

A determinist view underpins the discourse, one is which a national STEM agenda is represented as bringing economic benefits and general prosperity, thereby suggesting a simple pipeline from one to the other. Underpinning this discourse is a fear of falling behind other countries who share the same vision for a future dependent on STEM. Often this is accompanied by a sense of urgency and couched around the need for a national STEM strategy and action plan. The discourse expresses concern about the current state of school education to support the realisation of this envisioned future. It advocates for teachers to take a much greater role in achieving this future such as reversing the trend in declining student participation and performance levels in STEM.

STEM is important to education and to Australia's future. Schools do have an important role to play in ensuring that today's students are equipped with the knowledge and skills to gain jobs in the future. Many of these jobs are forecast in STEM. However, we need to be mindful that the vision of STEM presented in these national and state policies is a rather simplistic one. It is one for example, in which schools and teachers are represented as easily able to increase participation and performance in STEM. Complexities in this endeavour are downplayed or ignored. This simplistic view of STEM, becomes difficult to challenge. All too easily teachers and schools can be blamed when this discourse is not so easily realised in practice. As the STEM discourse continues to evolve, what is needed is a critical discourse, one that recognises the complexities in the STEM enterprise.

References

Ball, S. J. (1993). What is policy? Texts, trajectories and toolboxes. *The Australian Journal of Education Studies, 13*(2), 10–17. doi:10.1080/0159630930130203

Department of Education and Training. (2016). *STEM in the education state.* Retrieved from http://www.education.vic.gov.au/Documents/about/programs/learningdev/vicstem/STEM_EducationState_Plan.pdf

Education Council. (2015). *National STEM school education strategy: A comprehensive plan for Science, Technology, Engineering and Mathematics education in Australia.* Retrieved from http://www.educationcouncil.edu.au/site/DefaultSite/filesystem/documents/National%20STEM%20School%20Education%20Strategy.pdf

Fimyar, O. (2014). What is policy? In search of frameworks and definitions for non-western contexts. *Educate, 14*(3), 6–21.

Gee, J. (2005). *An introduction to discourse analysis: Theory and method.* New York, NY: Routledge.

Gee, J. (2011). *How to do discourse analysis: A toolkit.* New York, NY: Routledge.

Greckhamer, T., & Cilesiz, S. (2014). Rigor, transparency, evidence, and representation in discourse analysis: Challenges and recommendations. *International Journal of Qualitative Methods, 13*, 422–443.

Hewitt, S. (2009). *Discourse analysis and public policy research* (Centre for Rural Economy Discussion Paper Series No. 24). Retrieved from http://ippra.com/attachments/article/207/dp24Hewitt.pdf

Lankshear, C., Snyder, I., & Green, B. (2000). *Teachers and techno-literacy: Managing literacy, technology and learning in schools.* St Leonards: Allen & Unwin.

Lloyd, M. M. (2008). Uncertainty and certainty: The visions and roadmaps of ICT educational policy. *Computers in New Zealand Schools, 20*(3), 13–21.

Moyle, K. (2005, November 27–December 1). *Computing technologies in school education: Policies and standards and standard policies.* Paper presented at the Australian

Association for Research in Education (AARE) 2005, Parramatta, Sydney. Retrieved October 17, 2006, from http://www.aare.edu.au/05pap/moy05462.pdf

Office of the Chief Scientist. (2013). *Science, technology, engineering and mathematics in the national interest: A strategic approach*. Retrieved from http://www.chiefscientist.gov.au/wp-content/uploads/STEMstrategy290713-FINALweb.pdf

Office of the Chief Scientist. (2014). *Science, technology, engineering and mathematics: Australia's future*. Retrieved from http://www.chiefscientist.gov.au/wp-content/uploads/Australias-STEM-workforce_full-report.pdf

Selwyn, N. (2002). *Telling tales on technology: Qualitative studies on technology and education*. Hampshire: Ashgate Publishing Limited.

Selwyn, N. (2011). *Schools and schooling in the digital age: A critical analysis*. Abingdon: Routledge.

Van Dijk, T. A. (2015). Critical discourse analysis. In D. Tannen, H. E. Hamilton, & D. Schiffrin (Eds.), *The handbook of discourse analysis*. New York, NY: John Wiley & Sons.

CHAPTER 4

Implementing Virtual Reality in the Classroom: Envisaging Possibilities in STEM Education

Grant Cooper and Li Ping Thong

Abstract

With the advancement of immersive virtual reality (VR) there are various possibilities with the introduction of these technologies. Preparing students to effectively navigate, contribute to, and participate in virtual environments appears to be an important set of STEM-related competencies in the future. This chapter describes the *VR Education Model* (VEM), describing elements of this technology and its possible application in the classroom. One factor in student underachievement in STEM subjects may be a heavy reliance upon textual representations at the expense of more visuo spatial representations. Therefore, the use of VR may be particularly beneficial when representing and learning about STEM-related concepts. The authors envisage a number of scenarios that include but are not limited to the possibilities described in this chapter. The implementation of VR is discussed in terms of a broader STEM vision that meets the unique needs and priorities of each school.

Keywords

Virtual reality – immersive virtual reality – immersive virtual reality in education – STEM education – VR Education Model

1 The Fourth Industrial Revolution and STEM Education

The ever-increasing digitalisation of societies is predicted to impact various elements of our lives including, but not limited to, future job roles and growth, societal equality, health and security matters. It has been labelled as the Fourth Industrial Revolution, representing new ways in which digital technologies become embedded within societies (Schwab, 2017). Notable examples of the revolution include advancements in robotics, The Internet of Things, artificial intelligence, quantum computing, additive processing and autonomous

 | DOI:10.1163/9789004391413_005

vehicles. Against a backdrop of climate change, aging societies and globalisation, future societies will face new and considerable challenges not faced by previous generations (Heller, 2009). As stakeholders consider future possibilities, an important consideration is to think about the skills and knowledge that students will need in order to function in the future.

Globally, STEM (Science, Technology, Engineering, Maths) competencies have been heralded by governments, industry and others as crucial to growth, fostering innovation and global-competitiveness (Carter, 2017). As many countries shift economies and labour markets towards higher-skilled, knowledge and service-based industries (Powell & Snellman, 2004), building STEM-related competencies is viewed as a priority for many countries. A key driver of this motivation relates to concerns about significant disparities between supply and demand in the STEM labour pipeline. For instance, in Australia STEM employment is predicted to grow 50% faster than other jobs with significant increases in professional, scientific and health care roles (Hobbs, Cripps Clark, & Plant, 2018). It is estimated that changing 1 per cent of Australia's workforce into STEM-related roles would add $57.4 billion to GDP (PWC, 2015). As part of the response, education reform has been viewed as one way of addressing the significant mismatch between STEM labour market demand and domestic supply. The growing digitisation of societies means that digital literacies are positioned as an important element of students' STEM learning needs. It is predicted that by 2020, half of all Australians will need higher levels of digital literacies, including capacity to program, test software and build digital technologies (Regional Australia Institute, 2016).

2 Immersive Virtual Reality

As part of the growing digitisation of our communities, immersive virtual reality (shortened to virtual reality or VR herein) is viewed as a transformative tool, with the potential to impact various facets of our lives. Virtual reality can be broadly defined as an experience in which agents interact within a three-dimensional (3D) world with movement of their body, experiencing images and sounds (Sherman & Craig, 2002). This 3D environment is an attempt to replicate reality or the 'real world' in some way. Distinguishing VR from all preceding technology is the sense of immediacy and control, a feeling of 'being there' that comes from a changing visual display that is dependent on one's movements (Psotka, 1995). For example, to move an object in a VR environment, you may grab the object with your hands, lift it as you normally lift objects in the real world, and put it down where desired inside the virtual environment. Within these VR environments, the user can manipulate 'digital objects' in the same manner as they would manipulate them in the real world, as opposed to typing/pointing/clicking traditionally used in computer environments (Fällman, Backman, & Holmlund, 1999).

While the VR phrase was coined in the late 80's (Sherman & Craig, 2002), its popularity and use has grown recently in many countries around the world. In part, the catalyst for its growing popularity has been the falling cost of hardware and the creation of VR content from companies such as Google, Sony and Facebook. It is predicted that the VR market alone, including hardware, networks, software and content, will reach $692 billion (US) by 2025 and will be a trillion-dollar (US) industry by the year 2035 (Citigroup, 2016). While much of the discussion about VR technologies has focused on the impacts it is having on the video-game industry, it is predicted that it will impact many facets of our daily life in the not too distant future (Dede, 2009).

At the time of writing, the two main platforms of VR include desktop and mobile platforms. Through head-mounted displays connected to desktop computers, users are able to actively engage in high-fidelity virtual environments – for instance, interact with virtual objects with their hands, dynamically changing views and position within the virtual world through head and body movements, etc. (Bowman & McMahan, 2007). With a 360-degree display and head tracking, users are able to view through a pair of virtual eyes – no longer confined within standard frames or screen sizes often associated with other digital media devices (e.g. television, computer screens, mobile devices). More sophisticated desktop VR technologies offer higher fidelity, shorter latency (short delays in the visual update as users pivot their head), and wider field-of-views (how much could the users see within the head mounted display). In addition to head tracking, headphones are often used to utilise binaural and 3D audio to experience the digital soundscape of the environment – enabling users to not only listen to realistic environmental and ambient sounds, but also detect directional audio effects (e.g. listening to a bird chirping at the back, or a car driving past their left, etc.).

Mobile VRs often involve different input methods to interact with the environment – such as gaze-input (users turn their head and gaze at the interactive element within the scene to trigger an action) and game controllers. In desktop VRs like the Oculus Rift and HTC Vive, the experience is further enhanced with controller and positional tracking. With tracking capabilities on controllers, users are able to view digital representations of objects physically being held in their hands. The controllers could be displayed as 'ghost hands' – a common VR design pattern which enables users to see their virtual hands within the digital environment, and in some instances – even finger movements and gestures could be detected via the controllers and displayed within the virtual scene. With positional tracking, users are able to also physically walk within the vicinity of a targeted space, simulating a sense of space within the digital environment.

For the purpose of this paper, the authors of this paper are focusing on elements of this technology and its possible application in the classroom. The paper is structured into four sections. First, there is a discussion about existing research reporting the *application of VR in education*. Second, the authors consider *learning theories, VR and its application in STEM Education*. Third, elements of the technology and its possible application in the classroom are examined in the VR education model, inclusive of *experiencing, engagement, equitability* and *everywhere*. Last, there is a discussion about the possible integration of VR in schools, perhaps a helpful starting point for stakeholders to consider if or how the technology might align more broadly with a school's vision and priorities.

3 Application of VR in Education

Literature discussing the application of VR in education has existed for some time.

For instance, Fällman, Backman, and Holmlund (1999), examined the potential of VR to help students visualize abstract concepts, to take part in and interact with events that for reasons of distance, time, scale, or safety would not otherwise be conceivable. Others, such as Jonassen (1999), have identified a set of principles of constructivist learning environments that are relevant to VR. Those principles include the importance of:

1. Providing multiple versions of reality, thereby representing the natural complexity of the world
2. Focusing on knowledge construction rather than reproduction
3. Presenting authentic tasks
4. Fostering reflective practice
5. Facilitating context and content-dependent knowledge construction and
6. Supporting collaborative constructions of knowledge, rather than encouragement of competition among learners for recognition.

Research from the 1990's has previously discussed ideas and principles about the application of VR in education. This technology, however, has evolved considerably since the 90's, offering a host of new opportunities and challenges unimagined in previous research. At the time of writing, VR hardware creators include Google (Daydream), Sony (PlayStation VR), Samsung (Gear VR), HTC

(Vive) and Oculus (Rift/Go). The release of new hardware allows greater scope for software developers to design increasingly complex virtual environments (Lee & Wong, 2014). In addition to the hardware, there are developers across the globe making VR content including games, apps and animations. Mark Zuckerberg, the chief executive officer of Facebook, has claimed that VR will be the main social media platform in five to ten years, adding that technology changes in constant time waves and the next wave will be virtual reality (Baig, 2016). Consequently, hardware and software developments have implications for the kinds of possible learning experiences possible on the platform. Therefore, it is timely to consider the scope of VR technologies and its potential impact on teaching and learning. In order to consider some of the possibilities associated with the application of VR, the authors of this paper built on the work of Shuck and Aubusson (2010) to consider conceivable *"educational scenarios"* (p. 293). The authors of this chapter envisage a number of scenarios that include but are not limited to those described in this chapter. Some scenarios discussed in this chapter may be more likely to occur in relatively shorter timeframes than other, longer-term possibilities. The value of these scenarios lies in their ability to stimulate questions, rather than the accuracy of their predictions. Exploring possibilities regarding the increased integration of VR into teaching pedagogies is intended to give the reader a greater appreciation of the potential of this technology in an educational context, with a particular focus in scenarios related to learning STEM-related concepts.

4 Learning Theories, VR and its Application in STEM Education

Discourses about the process of learning commonly include behaviourist, cognitive constructivist and social constructivist paradigms. From a behaviourist viewpoint, the learner passively absorbs a pre-defined body of knowledge (Jarvis, Griffin, & Holford, 2007). Typically, repetition or 'drills' and positive reinforcement are pedagogies associated with behaviourist perspectives. Cognitive constructivists, such as Jean Piaget, rejected the behaviourist assumption that learning was the passive assimilation of knowledge (Bergin & Bergin, 2016). Instead, learners construct knowledge by creating and testing their own theories of the world, occurring through discrete stages in the cognitive development of children. Similarly, social constructivists such as Lev Vygotsky emphasised that learning is constructed socio-culturally; highlighting the importance of interactions with one another and the importance of environment in meaning making (Bergin & Bergin, 2016). While cognitive and social constructivists both agree that one needs to actively construct their learning, social constructivists view learning more as a social experience, not a process of behavioural development shaped by developmental stages. VR apps could be (and have been) designed with a range of underlying learning perspectives. For instance, a student spelling repetitive lists of words off a virtual blackboard might align quite closely with a behaviourist paradigm. Alternatively, VR software has the potential to promote social constructivist perspectives, by providing environments to actively rehearse, generalise, master and construct skills and knowledge. Within a VR environment, knowledge can be actively constructed through interaction with objects and scenarios (Sherman & Craig, 2002). Below, the authors of this chapter explore the elements of VR that have the potential to transform how students might learn while using the platform. The authors consider four important elements of VR in its application as an educational tool including *experiencing*, *engagement*, *equitability* and *everywhere*. We describe this framework as the *VR Education Model* (VEM).

The use of VR may be particularly beneficial when representing and learning about STEM-related concepts in particular. For instance, Gates (2018) emphasised the importance of multiple representations when teaching STEM-related subject matter, stating that one root cause in underachievement can be traced back to a heavy reliance upon textual representations at the expense of more visuospatial representations. The immersive potential of the platform means that it appears able to represent visuospatial elements more effectively than other previously used mediums.

5 Experiencing

An important element of VR in its application as an educational tool is the immersive potential of the platform. In VR environments, people commonly respond physically (e.g. facial reactions, movement of arms) and emotionally (e.g. shock, surprise) to situations even when they are aware that the phenomena is fictional and hence, the individual is commonly willing to suspend disbelief (De la Peña et al., 2010). Use of VR offers the potential to investigate distant locations, explore hidden phenomena, and manipulate otherwise immutable structures (Lee & Wong, 2014). For example, VR environments have the potential to allow students to virtually explore the surface of Mars or visit a historical period or site.

VR also offers students the opportunity of experiencing the same phenomena, but from the viewpoint of different stakeholders (e.g. climate change from a range of stakeholder views). This creates opportunities for students to experience and critique different digital identities. This could be either self-generated or an already established character within the environment. The VR headset essentially serves as a virtual extension of a person, in which they are viewing through the eyes of their digital identity. This outcome has the potential to change how individuals view themselves (and others) as a greater number of their self-concepts may be less bound to their immediate environment. Past research has indicated that immersion in virtual environments can be an effective learning experience by enabling situated learning, transferal of knowledge and enabling users to experience different perspectives (Dede, 2009). Mastery of concepts involved a complex process of building mental models about phenomena that represents intangible concepts and abstractions, however, students generally lacked real-life analogies on which to build these mental models (Bu & Schoen, 2011). VR technologies facilitate the process of building mental models by offering students the chance to immerse themselves in the phenomena.

6 Engagement

When considering its application as an educational tool, VR has the potential to be highly engaging for learners. Immersion in virtual learning environments also appears to be highly motivating for students, leading them to spend more time on the learning task (Winn, Windschitl, Fruland, & Lee, 2002). O'Brien and Toms (2008) regard engagement with technology as a subset of experiences, identifying a range of indicators including: attention, novelty, interest, control, feedback, and challenge; further noting that evidence of emotional (affect and motivation), sensory (aesthetics and interactivity) and spatiotemporal (perception of time, self/external awareness). When comparing the immersive and multi-sensory experience of VR environments, other linear forms of digital media (e.g. video, audio, websites) may not be as effective in sustaining students' engagement. As such, VR environments offer exciting possibilities during learning, offering virtual realms with the potential for interactive experiences. VR may be considered by some to appeal to students because of its novelty. While novelty may promote engagement, educators should focus on the pedagogical potential of the technology. In other words, educators are encouraged not to see VR as simply a gimmick in the classroom, but as a tool that has the potential to radically transform teaching and learning.

VR environments with pleasing aesthetics and challenging tasks can be intrinsically motivating and fun, possibly promoting engagement with the subject content. One notable example includes *HoloLab Champions* – a VR environment that situates users in a futuristic laboratory, where users could virtually mix chemicals using the same physical actions as they would in a chemistry lab and experiment with countless chemical combinations to view simulated reactions. The scaffolded learning in *HoloLab Champions* has the potential to spark students' curiosity in chemistry in a safe and engaging environment, enabling an enjoyable learning experience.

In another example, *Wonderful You* takes one on a journey of a human's first nine months in their mother's womb. According to the developers of the game, it … "plunges you into the expanding sensory world of your unborn self. Safe in the womb, you hear music in your dreams, you taste what your mother eats, you see sunlight and colour, your hands grasp what they touch" (BDH Immersive, 2017, p. 2). The potential to use an app such as this during a STEM class certainly offers educators exciting opportunities to excite and engage students that were previously not possible.

Both examples of *HoloLab Champions* and *Wonderful You* offer a glimpse of the potential of VR as a powerful tool for students' learning engagement in education. Both examples encourage students to actively participate with an emphasis on trial-and-error, gamified learning.

7 Equitability

When one considers VR and its application as educational tool, the concept of what the authors describe as equitability may be important element to consider. Within the scope of this paper, equitability broadly refers to stakeholder actions (e.g. student, teacher, caregiver) and environments that promote greater consistency in the education experiences of students. While equitability may overlap with notions of equity and equality, it has distinct elements. Discussed further in a moment, increases to consistency can be framed in both positive and negative ways. In this section, the authors present two scenarios that are intended to give the reader greater insight into VR's applications and its potential to promote equitability:

> Consider a state-of-the-art exercise-science lab. In the physical world, these facilities are more likely to be available in metropolitan areas to students who attend elite education providers. Cutting edge facilities can be expensive to build and maintain – typically such facilities are more viable in large educational institutions with economies of scale an important consideration. However, a VR exercise-science lab, closely replicating one found in a world-class university, has the potential to be accessed from anywhere with an internet connection. Maintenance costs of physical resources (e.g. cleaning, security) may also be lower. The technology, for instance, may be able to provide all students with the same virtual learning

environments (equality) and address barriers (e.g. geographical, economic) that may otherwise prevent some students from accessing these spaces (equity). In this scenario, VR promotes equitability because of the greater consistency in the learning environments students are accessing.

Other possible outcomes may be more complex than the previous example. For instance, there is also potential for students and others in the school community to present alternative and multiple-selves, interacting with digital content that has been collaboratively made – sharing the experiences of worlds beyond the physical (Churchill, Snowdon, & Munro, 2002). Students could potentially appear differently depending on their preferences, and at different stages of the day, change their digital self and identities. Some see this as an unprecedented era in human history, where the self can be transformed into anything imagined by the animator (Arazi, 2017). VR presents as a transformative technology in how students and institutions view and respond to sameness and differences in schools (and beyond). For instance, students may have the option to modify elements of their digital avatar in order to change how others responded to them in the school yard. If this possibility exists in the future, some students may 'go with the crowd' and choose features of their digital avatar considered to be on trend or preferable in some way. At the other end of this continuum, students may take the opportunity to be intentionally different from their peers, changing their avatar in ways that digress from the norm. If this turns out to be the case, it could be argued that VR offers greater autonomy to students in the way(s) they choose to represent themselves in school (and other environments). It may also be argued that VR has the potential to homogenise cohorts of students, turning them into groups of virtual avatars that could align with a hegemonic discourse of what is considered to be on trend or desirable in some way. If this turns out to be the case, the homogeneity of students' virtual avatars may lead to greater equitability in school systems. Students may be able to digitally select their level of sameness or difference (relative to peers/others), the possible impacts of this on equality and equity in educational contexts is potentially quite significant, complex and unknown. How institutions, educators and parents respond to such possibilities is also yet to be seen. Perhaps this section raises more questions than answers because VR has the potential to completely transform environments and stakeholder identities within these spaces.

8 Everywhere

VR technologies offer exciting possibilities in relation to where, when and how learning takes place. Collaborative Virtual Environments (CVEs) are online environments where people can meet and interact with other students, teachers and with digital content (e.g. 3D models, video files) (Churchill, Snowdon, & Munro, 2002). VR learning environments have the potential to allow students to problem solve in teams, potentially comprising of learners from all around the world (although different time zones and languages may impact the types of students in the CVE). While it could be argued that online teaching environments may achieve similar outcomes to CVEs, the opportunities to interact and manipulate 3D objects in real time with others arguably surpass the kinds of experiences currently possible. When one considers the possible implications of CVEs in terms of its implications for learners, it is potentially an '*everywhere*' learning medium, where learning can occur *anywhere*, at *anytime* with potential input from physical or virtual peers, teachers and facilitators, unbounded by the possible restrictions of other learning modalities. Considering the global potential of CVEs, VR has the potential to radically blur the lines between school and other domains of students' lives (e.g. home, sports, hobbies), allowing children to easily move between environments of their choosing from the one physical location.

VR technologies have the potential to totally transform teaching and learning (and daily life for many). As discussed, multiple representations of STEM-related concepts may be more effectively taught with pedagogies that allow students to explore visuospatial elements instead of a traditional overreliance on textual representations (Gates, 2018). VR platforms have the potential to immerse students in environments which allow them to represent, analyse, and virtually manipulate objects.

9 Potential Barriers to Overcome?

There may however, be a range of factors that may limit or inhibit the integration of VR technologies. Reasons might include, for instance, teacher self-efficacy, professional development opportunities, school leadership priorities and the amount of access to VR in schools. One important piece of the puzzle appears to be teachers and their professional learning. The implementation of virtual and mixed realities may be a considerable pedagogical shift for many in-service teachers. Stakeholders may need to invest considerably in comprehensive professional learning for in-service teachers. The costs associated with the implementation of VR technology may also be a salient consideration. The fiscal outlay for the technology could range on a spectrum from school to parent responsibility, or somewhere in between. The kind of VR platforms are another consideration. There may be a range of possibilities here, with current options including, for example, students bringing their smartphone to a headset provided by the school or dedicated VR headsets. Learning institutions commonly mirror societal trends including the adoption and use of digital technologies (Shuck & Aubusson, 2010). Hence, broader societal trends related to the acceptance and use of VR may be a salient factor that impacts the up-take of it in schools.

10 Considering the Use of VR in a School's STEM Education Vision

Keeping in mind the transformative potential of VR, the authors argue that it is important for educators to consider if and how this technology, including its implementation as a pedagogical tool, aligns more broadly with the school's STEM vision. Hobbs, Cripps Clark, and Plant (2018) emphasised the importance of establishing a STEM vision that is unique to each school's needs – "one that addresses the subtle and complex challenge of preparing 21st-century citizens within the constraints of a traditional school system and curriculum" (p. 139). A STEM vision may detail (1) the framing of STEM education in the school (e.g. subject oriented or interdisciplinary), (2) focuses on STEM teaching and learning pedagogies, (3) curriculum planning and assessment practices, (4) teacher learning and (5) community/industry engagement (Hobbs et al., 2018). If VR use increases as predicted, stakeholder discussion that focuses on the five elements of a STEM vision might be an effective starting point for considering the integration of this technology. As schools potentially shift from teaching in physical locations to multiple mixed reality environments, there are many possible impacts on teaching pedagogies, environments and school structures.

11 Conclusion

The ever-increasing digitalisation of societies has a number of potential and profound impacts. As part of this predicted change, students may need to effectively navigate a number of virtual (or mixed reality) environments in the future. Preparing students to effectively navigate, contribute to and participate in virtual environments appears to be an important future set of STEM-related skills and knowledge. VR platforms also act as a potentially transformative educational tool. This chapter describes the *VR Education Model* (VEM), describing elements of this technology and its possible application in the classroom. The authors consider four elements of VR in its application as an educational tool including *experiencing*, *engagement*, *equitability* and *everywhere*. While future trends in education can be difficult to predict, the authors have described a number of possible scenarios that include but are not limited to the possibilities described in this chapter. The implementation of VR is discussed in terms of a broader STEM vision that meets the unique needs and priorities of each school. There are complex considerations about the integration of VR that educators may need to respond to if future predictions of how the technology will evolve are accurate.

References

Arazi, K. (2017). The virtues of virtual reality. *Medium*. Retrieved January 23, 2018, from https://medium.com/@kimarazi/the-virtues-of-virtual-reality-f2a67daf2cb6

Baig, E. (2016). Mark Zuckerberg: Virtual reality can become the most social platform [online]. *USA Today*. Retrieved July 8, 2017, from https://www.usatoday.com/story/tech/columnist/baig/2016/02/21/mark-zuckerberg-vr-can-become-most-social-platform/80706338/

Bergin, C., & Bergin, D. (2016). *Child and adolescent development in your classroom, topical approach*. Melbourne: Cengage.

Bowman, D., & McMahan, R. (2007). Virtual reality: How much immersion is enough? *Computer, 40*(7), 36–43. Retrieved from http://dx.doi.org/10.1109/mc.2007.257

Bu, L., & Schoen, R. (2011). *Model-centered learning*. Rotterdam, The Netherlands: Sense Publishers.

Carter, L. (2017). Neoliberalism and STEM education: Some Australian policy discourse. *Canadian Journal of Science, Mathematics and Technology Education, 17*(4), 247–257. Retrieved from http://dx.doi.org/10.1080/14926156.2017.1380868

Churchill, E., Snowdon, D., & Munro, A. (2002). *Collaborative virtual environments*. London: Springer.

Citigroup. (2016). *Virtual and augmented reality*. Retrieved January 23, 2018, from https://www.citi.com/commercialbank/insights/assets/docs/virtual-and-augmented-reality.pdf

Dede, C. (2009). *Introduction to virtual reality in education*. Retrieved 19 January, 2018, from http://earthlab.uoi.gr/theste/index.php/theste/article/view/20

De la Peña, N., Weil, P., Llobera, J., Spanlang, B., Friedman, D., Sanchez-Vives, M., & Slater, M. (2010). Immersive journalism: Immersive virtual reality for the first-person experience of news. *Presence: Teleoperators and Virtual Environments, 19*(4), 291–301. Retrieved from http://dx.doi.org/10.1162/pres_a_00005

Fällman, D., Backman, A., & Holmlund, K. (1999). *VR in education: An introduction to multisensory constructivist learning environments*. Retrieved July 8, 2017, from https://www.semanticscholar.org/paper/VR-in-Education-An-Introduction-to-Multisensory-CF%C3%A4llman/f71bddbba8fe4d040433b7b10587978af618f7a9

Gates, P. (2018). The importance of diagrams, graphics and other visual representations in STEM teaching. In R. Jorgensen & K. Larkin (Eds.), *STEM education in the junior secondary*. Singapore: Springer.

Heller, P. (2009). *Who will pay?* Washington, DC: International Monetary Fund.

Hobbs, L., Cripps Clark, J., & Plant, B. (2018). Successful students – STEM program: Teacher learning through a multifaceted vision for stem education. In R. Jorgensen & K. Larkin (Eds.), *STEM education in the junior secondary*. Singapore: Springer.

Jarvis, P., Griffin, C., & Holford, J. (2007). *The theory & practice of learning*. London: Routledge.

Jonassen, D. H. (1999). Designing constructivist learning environments. In C. M. Reigeluth (Ed.), *Instructional design theories and models: A new paradigm of instructional theory* (Vol. II, pp. 215–239). Mahwah, NJ: Lawrence Erlbaum Associates.

Lee, E., & Wong, K. (2014). Learning with desktop virtual reality: Low spatial ability learners are more positively affected. *Computers & Education, 79*, 49–58.

O'Brien, H. L., & Toms, E. G. (2008). What is user engagement? A conceptual framework for defining user engagement with technology. *Journal of the Association for Information Science and Technology, 59*(6), 938–955.

Powell, W., & Snellman, K. (2004). The knowledge economy. *Annual Review of Sociology, 30*(1), 199–220. doi:10.1146/annurev.soc.29.010202.100037

Psotka, J. (1995). Immersive training systems: Virtual reality and education and training. *Instructional Science, 23*(5–6), 405–431. Retrieved from http://dx.doi.org/10.1007/bf00896880

PWC. (2015). *A smart move: Future-proofing Australia's workforce by growing skills in Science, Technology, Engineering and Maths (STEM)*. Sydney: Pricewaterhouse Coopers.

Regional Australia Institute. (2016). Thefutureofwork.net.au. Retrieved January 22, 2018, from http://www.thefutureofwork.net.au/Download/File/TheFutureofWork_ReportNovember2016.pdf

Schuck, S., & Aubusson, P. (2010). Educational scenarios for digital futures. *Learning, Media and Technology, 35*(3), 293–305.

Schwab, K. (2017). *The fourth industrial revolution*. London: Penguin.

Sherman, W., & Craig, A. (2002). *Understanding virtual reality: Interface, application, and design*. Burlington, MA: Morgan Kaufmann.

Winn, W. D., Windschitl, M., Fruland, R., & Lee, Y. (2002). When does immersion in a virtual environment help students construct understanding? In P. Bell & R. Stevens (Eds.), *Proceedings of the international conference of the learning societies*. Mahwah, NJ: Lawrence Erlbaum Associates.

CHAPTER 5

Multiplicative Thinking: A Necessary STEM Foundation

Dianne Siemon, Natalie Banks and Shalveena Prasad

Abstract

Across the science, technology and engineering fields there is very little of any substance that can be achieved without the capacity to recognise, represent and reason about relationships between quantities, that is, to think multiplicatively. However, recent research has found that at least 25% and up to 55% of Australian Year 8 students are not demonstrating a capacity for multiplicative thinking. This helps explain the decline in the relative performance of Australian students on international assessments of mathematics and the significant decline in the proportion of Year 12 students undertaking the more advanced mathematics courses. But the data also reveal significant inequities in that students from low socioeconomic communities are far more likely to be represented in the 45 to 55% range than students from higher socioeconomic backgrounds who are more likely to be represented in the 25 to 35% range. This situation is untenable where the fastest growing employment opportunities require some form of STEM qualification. The chapter presents evidence from two large scale research projects to make a case for focussing on identifying and responding appropriately to students' learning needs in relation to multiplicative thinking as a key priority in STEM education.

Keywords

Multiplicative thinking – learning progressions – targeted teaching – formative assessment – quantitative reasoning

1 Introduction

Numerous industry and government sponsored reports have pointed to the threat to Australia's economy posed by lack of access to a suitably qualified STEM workforce. That is, to workers with the appropriate knowledge, skills and

 | DOI:10.1163/9789004391413_006

experience in science, technology, engineering and mathematics (Australian Industry Group, 2015; Business Council of Australia, 2015; Pricewaterhouse-Cooper, 2015; Office of the Chief Scientist, 2012, 2013).

> The global economy is changing. New technologies and smart companies lead. New industries and new sources of wealth are emerging. New skills are required for workers at all levels. … At the core of almost every agenda is a focus on STEM: science, technology, engineering and mathematics. (Office of the Chief Scientist, 2014, p. 5)

This situation is compounded by the sharp decline in the proportion of students undertaking higher level mathematics and science subjects in the final years of schooling (Australian Industry Group, 2015) and the significant decline in Australian students' performance on international assessments of mathematical literacy relative to other countries (e.g., Thomson, De Bortoli, & Underwood, 2017; Thomson, Wernet, O'Grady, & Rodriguez, 2017). Of particular concern is data from the 2012 Programme for International Student Assessment (PISA) that reveal 20% of Australian 15-year olds fell short of the minimum international proficiency standard (Level 2), the level at which students are beginning to demonstrate the competencies needed to participate effectively and productively in society (Thomson, De Bortoli, & Buckley, 2013). The 2012 PISA results also revealed that student performance is strongly and positively related to socioeconomic background, Indigenous students continue to achieve on average at a level approximately two years below their non-Indigenous peers, and approximately one-third of the Australian females reported "they did not think mathematics was important for later study compared to one-fifth of the males" (p. 223). A result reflected in the fact that women "continue to leave STEM in unacceptably high numbers at secondary, tertiary and early-career level" (Office of the Chief Scientist, 2014, p. 21). These inequities are untenable where an estimated 75% of the fastest growing employment opportunities require "significant science or maths training" (Becker & Park, 2011, p. 23). Not surprisingly, these issues have prompted business leaders, politicians and educators to call for a focus on STEM education. A call that is so widespread and popular that Sanders (2009) has referred to it as STEMmania.

2 STEM Education – A Contested Notion

Countries that have "a strong track record in innovation also tend to have a strong commitment to STEM education and as a result a strong pipeline of

STEM workers" (Australian Industry Group, 2015, p. 16). However, what this means in practice varies by sector and context. For instance, at the university level, STEM education may involve an extended, in-depth study of a particular aspect of mathematics or a multi-disciplinary approach to issues of sustainability. At the enterprise level, it might involve on-the-job training in the use of a new technology. At the school level STEM education involves a focus on the big ideas and strategies of the underpinning disciplines as well as the application of those ideas and strategies in rich, multi-disciplinary contexts.

> STEM education is a term used to refer collectively to the teaching of the disciplines within its umbrella – science, technology, engineering and mathematics – and also to a cross-disciplinary approach to teaching that increases student interest in STEM-related fields and improves students' problem solving and critical analysis skills. (Rosicka, 2016, p. 4)

While there are many that advocate the second of these two approaches for STEM education in schools (e.g., Furner & Kumar, 2007; Rosicka, 2016; Sanders, 2009) even defining STEM education in terms of "approaches that explore teaching and learning between/among any two or more of the STEM subject areas, and/or between a STEM subject and one or more other school subjects" (Sanders, 2009, p. 21), there are others who urge caution (e.g., Larson, 2017; Venville, Rennie, & Wallace, 2009). Larson (2017) for instance, agrees with Bybee (2013), a respected science educator, that the "purpose of STEM education is to develop the content and practices that characterize the respective STEM disciplines" (Bybee, 2013, p. 4).

> If in the "STEM program" the mathematics isn't on grade level, or if the mathematics isn't addressed conceptually but rather as a procedural tool to solve various disjointed applications, or if the mathematics is not developed within a coherent mathematical learning progression, then the "STEM program" fails the fundamental design principle. … to develop the content and practices that characterize effective mathematics programs while maintaining the integrity of the mathematics. (Larson, 2017)

Those who argue for integrated approaches to STEM in schools tend to do so on the grounds of process-oriented outcomes such as critical thinking, analyzing, explaining, generalizing, applying, evaluating, refining, and communicating (e.g., Rosicka, 2016). While these are highly desirable goals and there is evidence to suggest that integrated approaches provide "opportunities for more relevant, less fragmented, and more stimulating experiences for learners" (Furner & Kumar, 2007, p. 186), there is little evidence that integrated approaches lead to a deep understanding of important mathematical ideas and the connections between them (Larson, 2017).

Mathematics is fundamental to science, engineering and technology (Wai, Lubinski, & Benbow, 2009), and while it is undoubtedly true that integrated approaches provide opportunities to apply and extend the mathematics that is already known, it is incumbent on schools and mathematics educators to ensure that students acquire the knowledge and skills they will need to successfully participate in and contribute to the solution of more challenging problems. If the mathematics is only prompted by students on a 'need to know' basis arising from an integrated STEM activity, there is a real risk that mathematics will be 'tacked on' and/or reduced to a set of narrow, disconnected skills and procedures. For example, an inquiry at Year 7 on the design and construction of a school kitchen garden provides opportunities to question, plan, design, measure, calculate, reason, represent, evaluate, communicate, explain and justify. It also has the potential to explore/address several curriculum content descriptors at this level as shown in Table 5.1 (note, this is indicative only, it is not meant to be exhaustive). However, the extent to which the Kitchen Garden inquiry fulfils this potential is entirely dependent on how knowledgeable and well prepared the teacher is to guide the inquiry in purposeful and productive ways (Badley, 2009; Kirschner, Sweller, & Clark, 2006). But it also depends on the students' prior mathematical knowledge and experience even if they are able to generate the sort of questions and plans that might prompt a consideration of the potential content. If the mathematics is only considered 'in passing' as a 'means to an end,' it is highly likely that the teaching focus will be procedural rather than conceptual and isolated and context-specific rather than connected and generalised.

An example that might arise in this context on a 'need to know' basis is how to solve a proportion problem such as, 'A garden bed design is 1.2 meters wide, 2 meters long and 45 cm high. Find the cost of filling it with soil if garden soil is sold in 25 Litre bags which cost $12 each.'

Unfortunately, one of the pedagogical issues of working with real world problems is that they are often 'messy,' that is, they involve multiple steps and quantities and relationships that in themselves pose problems for many students at this level (Lamon, 2007; Siemon, 2016; Siemon & Virgona, 2001). Quite apart from the knowledge and understanding needed to translate the metric units involved (place value and metric relationships) and undertake the necessary calculations (multiplication and division) to find out how much soil is needed, on completion of the initial steps students are left with something like 'if 1 fortieth of a cubic metre costs $12, how much will 1.08 cubic meters cost?'

TABLE 5.1 Indicative potential of kitchen garden inquiry to address year 7 curriculum expectations

Science	Engineering & technology	Mathematics
Biological Sciences – plant types, beneficial animals or insects ... Chemical Sciences – soil Ph, nutrients, fertilizer (ethics) Physical Sciences – light, energy, location, wind levels Nature and Development of Science – cultural, historical agricultural practices Use and Influence of Science – advances (and disasters) in agricultural science, sustainability (composting, water consumption) Inquiry skills – questioning, predicting, planning, conducting, processing, analysing, evaluating, communicating	Optimal design given space, location, cost Scale models, pilot experiments/trials Choice of fit-for-purpose materials, considerations (e.g., cost, environmental impact) Use of digital tools (e.g., computer drawing programs, calculators, spreadsheets)	Solve problems involving all four operations and whole numbers (Year 6) Add, subtract (Year 6), multiply and divide decimal fractions Recognise and solve problems involving simple ratios Investigate and calculate 'best buys' Establish the formulas of areas of rectangles, triangles and parallelograms Calculate the volume of rectangular prisms Draw different views of solids and prisms formed from a combination of prisms Investigate conditions for two lines to be parallel Construct and compare a range of data displays

In this situation, it would be quite understandable if a teacher scaffolded this by asking, 'What would 1 cubic meter cost? ($480), What do we need to do next? (multiply $480 by 1.08).' This addresses the immediate problem but it does so in a context-specific way that does not invite a consideration of missing value problems more generally or make connections to the underpinning big ideas of place value, multiplicative thinking, partitioning and proportional reasoning (Siemon, Bleckly, & Neil, 2012).

Dealing with the mathematics involved in this way is hardly conducive to developing a deep, well-connected knowledge base for mathematics that supports further learning and thereby engagement with STEM-related studies.

Nor is it likely to generate the sort of knowledge and confidence that everyone needs to make sense of the mathematics they encounter in their own lives such as working out which of two competing mobile phone plans is best suited to their needs. Clearly, a quality STEM education incorporates both approaches, that is, a coherent, well-planned sequence of mathematics instruction focused on developing the key ideas and strategies and the connections between them (e.g., Larson, 2017), and the opportunity to apply this knowledge in rich, integrated settings that require collaborative endeavor and exercise process skills (e.g., Rosick, 2016).

An important consideration here, is that the mathematics needs to be taught well. That is, it needs to reflect evidenced-based teaching practices that build on the knowledge and interests of students in purposeful and effective ways and value communication, collaborative problem solving and critical thinking (e.g., Boaler, 2008; Sullivan, 2011). In Australia, this would reflect a program that addresses all four of the proficiencies; described in the curriculum as understanding, fluency, problem solving, and reasoning (Australian Curriculum Assessment & Reporting Authority, 2016); that was not dominated by a focus on learning isolated facts and procedures but included opportunities to consider rich, cross-curriculum activities such as *Baby in the Car* and *Radioactivity*,[1] both of which require the application of multiplicative and critical thinking skills in STEM-related contexts. However, a quality mathematics education program also needs to recognise and respond appropriately to what students find difficult and prevents their further learning and their ability to apply what they know in integrated contexts. This has never been more important as Mike Lefkowitz noted in a Mind Research Institute blog in January 2014

> A solid foundation in mathematics and science develops and hones the skills of posing hypotheses, designing experiments and controls, analysing data, recognizing patterns, seeking evidence, conclusions and proof, solving problems and seeking absolutes, while being open to new information. Studying mathematics not only will develop more engineers and scientists, but also produce more citizens who can learn and think creatively and critically, no matter their career fields. The workforce of tomorrow, in all fields, will demand it. (p. 3)

Indeed, as Lefkowitz (2015) and Australia's Chief Scientist, Dr Alan Finkel (2017) – both very experienced scientists – have noted, the M in STEM is the most important element, there is no STEM without M. A sentiment that is echoed in an Australian Academy of Science (2016) report on the future of the mathematical sciences in Australia.

> Our world is changing rapidly. It is becoming increasingly technological and significantly more mathematical. If Australia is to prosper in such circumstances it will need:
> - Citizens who have been taught mathematics and statistics well while at school.
> - A steady flow of well-taught university graduates with advanced mathematical sciences skills.
> - A vibrant community of mathematicians and statisticians who are advancing the frontiers of the discipline for the broader benefit of Australian society. (p. 37)

However, the results of international assessments such as PISA 2012 suggest we could be doing better, particularly when more than 42% of Australian 15-year olds are below Level 3, which is the minimum nationally agreed standard (Thomson, De Bortoli, & Buckley, 2013). One of the reasons for this is, as exemplified in the example given above, is the known difficulties middle-year students experience with multiplicative thinking and rational number (Harel & Confrey, 1994, Lamon, 2007; Siemon & Virgona, 2001). The next section considers the evidence from two large-scale research projects that point to what students find difficult and what can be done to mitigate the risk of this situation continuing.

3 The Evidence

As background to the two large-scale studies to be reported here, it is necessary to briefly describe an earlier large-scale study, the *Middle Years Numeracy Research Project*[2] (MYNRP), which used rich assessment tasks to evaluate Year 5 to 9 students' capacity to use mathematics to solve unfamiliar problems and to explain and justify their thinking. Initial data were collected from a structured sample of just under 7000 Year 5 to 9 students from 27 primary schools and 20 secondary schools across Victoria in November 1999. Key findings included:
- there was as much difference in student numeracy performance within each year level as between year levels and that this difference was equivalent to 7 years of schooling;
- there was considerable within school variation indicating that individual teachers made a difference to student performance;
- that the needs of many students, particularly those who might be judged to be 'at risk' or 'left behind,' were not being met; and

- that irrespective of context, differences in performance were almost entirely due to an inadequate understanding of larger whole numbers, multiplication and division, fractions, decimals, and proportion, and a reluctance/inability to explain/justify solutions (Siemon & Virgona, 2001).

Typically addressed as isolated 'topics' with little/no reference to one another (Siemon, Bleckly, & Neal, 2012), these aspects of mathematics content were recognised by Vergnaud (1994) as constituting a Multiplicative Conceptual Field. That is, a framework of complex, inter-related ideas and strategies, which for the purposes of the MYNRP was described in terms of *multiplicative thinking*.

The results of the MYNRP show that while most students were able to solve multiplication problems involving relatively small whole numbers, they relied on additive strategies to solve more complex multiplicative problems involving larger whole numbers, rational numbers, and/or situations not easily modelled in terms of repeated addition. This suggested that the transition from additive to multiplicative thinking was nowhere near as smooth or as straightforward as most curriculum documents seemed to imply, and that access to multiplicative thinking represented a real and persistent barrier to many students' mathematical progress in the middle years of schooling (Siemon & Virgona, 2001).

This observation was supported by the literature available at the time. For example, there was a considerable body of research pointing to the difficulties students experience with multiplication and division (Anghileri, 1999; Mulligan & Mitchelmore, 1997), and the relatively long period of time needed to develop these ideas (Clark & Kamii, 1996; Sullivan, Clarke, Cheeseman, & Mulligan, 2001). Students' difficulties with rational number and proportional reasoning had also been well documented (for example, Baturo, 1997; Harel & Confrey, 1994; Lamon, 1996; Misailidou & Williams, 2003). And there was a growing body of research documenting the link between multiplicative thinking and rational number ideas (Baturo, 1997; Harel & Confrey, 1994); multiplicative thinking and spatial ideas (Battista, 1999), and the importance of both as a basis for understanding algebra (Gray & Tall, 1994). While this work contributed to a better understanding of the 'big ideas' involved, very little was specifically concerned with how these ideas relate to one another and which aspects might be needed when, to support new learning both within and between these different domains of multiplicative thinking. Moreover, very little of this work was represented in a form and language that was accessible to teachers or directly translated to practice in the middle years of schooling (Siemon & Virgona, 2001).

Simon's (1995) idea of constructing hypothetical learning trajectories (HLTs) as mini-theories of student learning in particular domains offered a useful means of addressing these problems as HLTs provide an accessible framework for identifying where students 'are at' and starting points for teaching. In Australia, learning trajectories have tended to take the form of *learning and assessment frameworks* which have been developed and validated in terms of several discrete domains such as counting, place-value, and addition in the early years of schooling (Clarke, Sullivan, Cheeseman, & Clarke, 2000). These are typically developed based on large-scale interview data and represented in a form that is accessible to teachers.

The *Early Years Numeracy Research Project* also found that where teachers were supported to identify and interpret student learning needs in terms of these frameworks, they were more informed about where to start teaching, and better able to scaffold their students' mathematical learning (Clarke, 2001). While learning and assessment frameworks for multiplication and division had been developed for the early years of schooling, the evidence suggested that very few students in Years P to 3 are at the point of abstracting multiplicative thinking, that is, able to work confidently and efficiently with multiplicative thinking in the absence of physical models (Mulligan & Mitchelmore, 1997; Sullivan et al., 2001). This suggested that developing an evidence-based, learning and assessment framework for the key ideas involved in multiplicative thinking that went beyond the early years and teachers could use to identify student learning needs and plan targeted teaching interventions, was likely to contribute to enhanced learning outcomes for students in the middle years.

4 The Scaffolding Numeracy in the Middle Years (SNMY) Project[3]

The results of the MYNRP and subsequent review of the literature led to the *Scaffolding Numeracy in the Middle Years* (SNMY) project, the aim of which was to which was to explore the development of multiplicative thinking in Years 4 to 8 with a view to developing an evidence-based learning and assessment framework that could be used to support a more targeted approach to the development of multiplicative thinking in the middle years of schooling (Siemon, Breed, Dole, Izard, & Virgona, 2006).

Although some aspects of multiplicative thinking are available to young children, multiplicative thinking is substantially more complex than additive thinking and may take many years to achieve (Lamon, 2007; Vergnaud, 1988). This is because multiplicative thinking involves recognising and working with relationships between quantities and it is concerned with processes such as replicating, shrinking, enlarging, and exponentiating, which are fundamentally more complex than the more obvious processes of aggregation and disaggregation associated with additive thinking and the everyday use of whole numbers.

For the purposes of the *Scaffolding Numeracy in the Middle Years* Linkage Project (SNMY, 2003–2006), multiplicative thinking was viewed in terms of:

- a capacity to work flexibly and efficiently with an extended range of numbers (for example, larger whole numbers, decimals, common fractions, ratio, and per cent),
- an ability to recognise and solve a range of problems involving multiplication or division including direct and indirect proportion, and
- the means to represent and communicate this effectively in a variety of ways (for example, words, diagrams, symbolic expressions, and written algorithms) (Siemon et al., 2006).

The methodology involved three overlapping phases. Phase 1 identified a broad hypothetical learning trajectory (Simon, 1995) which formed the basis of a Draft Learning and Assessment Framework for Multiplicative Thinking (LAF). Phase 2 involved the design, trial and subsequent use of a range of rich assessment tasks developed to evaluate various aspects of multiplicative thinking. The tasks, and their associated scoring rubrics, were variously used at the beginning and end of the project to inform the development of the LAF using Rasch modelling (Bond & Fox, 2001). Phase 3 involved research school teachers and members of the research team in an eighteen-month action research study that progressively explored a range of targeted teaching interventions (Learning Plans) aimed at scaffolding student learning in terms of the LAF.

The study was conducted in six school clusters, 4 in Victoria (2 metropolitan and 2 regional) and 2 in Tasmania (both metropolitan). Each school cluster involved a secondary school and at least three primary schools. Just over 1500 Year 4 to 8 students and their teachers from the three *research* school clusters were involved in Phases 2 and 3 of the project. A similar group of Year 4 to 8 students from the three *reference* school clusters was involved in Phase 2 only. In all, 40 schools were involved in the SNMY Project.

The assessment options comprised a longer task incorporating multiple items of increasing complexity in the same context (e.g., an 11-item restaurant seating problem that progressively explored the relationship between the number of tables and the number of people who could be seated) and five to seven shorter tasks each of which incorporated two or three items. An example of a short task and its associated scoring rubric is given in Figure 5.1.

Responses to the initial round of assessment were received from just under 3200 Year 4 to 8 students from research and reference schools in March 2004. The resulting data were analysed using the Rasch partial credit model (Masters, 1982) which allows both students' performances and item difficulties to be measured using the same log-odds unit (the logit) and placed on an interval scale. Multiple assessment options were used in the initial round to

ADVENTURE CAMP ...

Camp Reefton offers 4 activities. Everyone has a go at each activity early in the week. On Thursday afternoon students can choose the activity that they would like to do again.

The table shows how many students chose each activity at the Year 5 camp and how many chose each activity at the Year 7 camp a week later.

	Rock Wall	Canoeing	Archery	Ropes Course
Year 5	15	18	24	18
Year 7	19	21	38	22

Camp Reefton Thursday Activities

a. What can you say about the choices of Year 5 and Year 7 students?

b. The Camp Director said that canoeing was more popular with the Year 5 students than the Year 7 students. Do you agree with the Director's statement? **Use as much mathematics as you can to support your answer.**

ADVENTURE CAMP ...

TASK:	RESPONSE:	SCORE
a.	No response or incorrect or irrelevant statement	0
	One or two relatively simple observations based on numbers alone (e.g., "Archery was the most popular activity for both Year 5 and Year 7 students," "More Year 7 students liked the rock wall than Year 5 students")	1
	At least one observation which recognises the difference in total numbers (e.g., "Although more Year 7s actually chose the ropes course than Year 5, there were less Year 5 students, so it is hard to say")	2
b.	No response	0
	Incorrect (No), argument based on numbers alone (e.g., "There were 21 Year 7s and only 18 Year 5s")	1
	Correct (Yes), but little/no working or explanation to support conclusion	2
	Correct (Yes), working and/or explanation indicates that numbers need to be considered in relation to respective totals (e.g., "18 out of 75 is more than 21 out of 100"), but no formal use of fractions or percent or further argument to justify conclusion	3
	Correct (Yes), working and/or explanation uses comparable fractions or percents to justify conclusion (e.g., "For Year 7 it is 21%. For Year 5s, it is 24% because 18/75 = 6/25 = 24/100 = 24%")	4

FIGURE 5.1 An example of a short task and its associated scoring rubric from the SNMY project 2003–2006

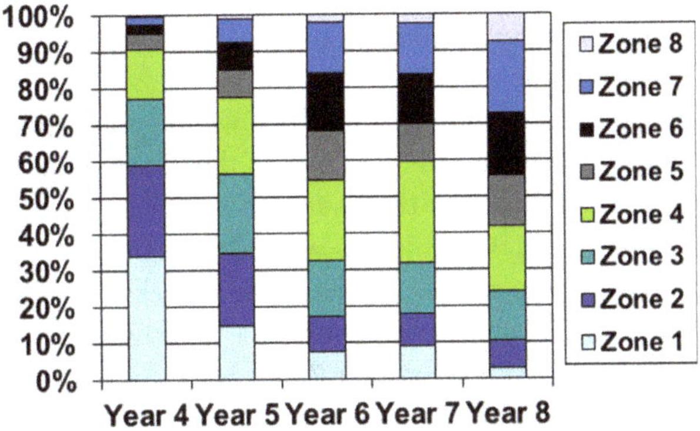

FIGURE 5.2 Proportion of students by LAF zone and year level, initial phase SNMY project, 2004 (n = 3169)

test item validity and inform the development of the LAF. The LAF identified as a result of this process comprised eight incremental Zones ranging from additive, count-all strategies (Zone 1) to the sophisticated use of proportional reasoning (Zone 8) with multiplicative thinking not evident on a consistent basis until Zone 4 (Siemon & Breed, 2006). Figure 5.2 shows the relative proportion of students at each year level in each zone of the LAF based on their responses to the initial assessment in 2004.

Given that, in curriculum terms, Zone 1 roughly corresponds to what might be expected by Years 1 and 2 with respect to multiplicative thinking and Zone 8 roughly corresponds to what might be expected by Years 8 and 9, the results confirm the findings of the MYNRP. That is, that there is a seven to eight-year range in student mathematics achievement in each year level that is almost entirely explained by the extent to which students have access to multiplicative thinking.

4.1 *Targeted Teaching*

The value of using assessment data to inform and improve teaching, generally referred to as formative assessment or assessment for learning, is widely recognised (e.g., Ball, 1993; Black & Wiliam, 1998; Callingham & Griffin, 2000; Clark, 2001; Earl & Katz, 2006). In phase 3 of the SNMY project, research school teachers worked with research team members to design Zone-specific activities based on the LAF that could be used to progress the learning of students from one Zone to the next. Referred to originally as *assessment-guided instruction*, the use of the Zone-specific activities came to be referred to as *targeted teaching* in the latter part of the SNMY project (Siemon et al., 2006) to distinguish the long-term, multi-faceted nature of the interventions needed to scaffold students' multiplicative thinking, from the equally valid but short-term or spontaneous teaching decisions that might be informed by a pre-test on subtraction or an informal classroom quiz.

The efficacy and long-term nature of targeted teaching was demonstrated by Breed (2011) who undertook a doctoral study as part of the SNMY project. Nine Year 6 students identified in Zone 1 of the LAF in 2004 participated in an 18-week intervention in mid-2005. The students worked with the teacher in small groups using manipulatives, games, discussion and weekly written reflections using the LAF as a guide. When re-assessed three months after the intervention, all nine students shifted at least four Zones with the majority shifting five Zones to be age and grade appropriate.

4.2 *Results*

The final SNMY assessment round was conducted in November 2005. Data were collected from 3350 Year 4 to 8 students and analysed using the Rasch partial credit model (Masters, 1982). The results confirmed the sequence of ideas and strategies in the Draft LAF and established the validity of the assessment tasks (Siemon et al., 2006). As the initial and final assessments were conducted in different school years, cohort comparisons rather than matched pairs were used to explore the impact of targeted teaching on student learning (e.g., growth from Year 4 to year 5 in research schools was compared to growth from Year 4 to Year 5 in reference schools). Overall, medium to large effect sizes (in the range 0.45 to 0.75 or more) as described by Cohen (1969) were found in research schools compared to small to medium effect sizes (in the range of 0.2 to 0.5) in reference schools. While the results varied between research schools, overall, they show that teaching targeted to identified student learning needs was effective in improving students' multiplicative thinking (Siemon et al., 2006).

5 The Reframing Mathematical Futures Priority Project (RMF-P)[4]

There are many reasons why Australian students choose not to pursue STEM-related studies in the senior years of schooling but, as indicated above, a major contributing factor is the seven to eight-year range in students' access to multiplicative thinking in the Years 4 to 9, which is needed to solve more difficult problems involving rational numbers and proportional reasoning and to successfully engage with algebraic, spatial and statistical reasoning (Siemon, 2013, 2016).

Since their publication on the Victorian Department of Education website in 2006, the SNMY assessment materials and the LAF have been used widely with upper primary students in Victoria, and to a lesser extent in Tasmania, South Australia, Northern Territory and Queensland. For a variety of reasons their use in secondary schools has not been as widespread. However, where they have been used, there is evidence to suggest that a targeted teaching approach based on the LAF is effective in improving mathematics outcomes for students in Years 7 to 9 (e.g., Reilly & Parsons, 2011).

Two key objectives of the Australian Mathematics and Science Partnership Program[5] (AMSPP) were to build the theoretical and pedagogical skills of school teachers to deliver maths and science subjects and increase the number of students undertaking maths and science subjects to Year 12. As Priority projects needed to be 'road ready," the *Reframing Mathematical Futures Priority* project (RMF-P) aimed at investigating how the SNMY materials could be used more effectively in Years 7 to 9 to improve student's multiplicative thinking and thereby increase their opportunities of accessing further STEM-related studies.

A condition of funding was that the project had a national focus and involved students from lower socioeconomic backgrounds. To this end, and because time was of the essence, State or Territory based education systems that were aware of the SNMY materials were approached to be partners in the project. Their key role at the outset was to identify schools that satisfied the funding condition (up to 6 in each participating State or Territory) where there was a willingness and capacity to provide some time release (preferably a day/week) for an existing teacher to participate in the project as an SNMY specialist. Project funding was provided to support the Specialists attend professional learning days (4 to 5 days), provide some time release (30 days/school) for other teachers (at least 2) to engage with the Specialist in school time, and support visits (at least 2) by members of the research team. A total of 28 secondary or middle schools agreed to participate in the project on this basis.

Teacher surveys undertaken as part of the SNMY project revealed that variations in research school results were largely explained by variations in the level of support provided by school leadership team and the knowledge and confidence of the teachers concerned in relation to multiplicative thinking (Siemon et al., 2006). As a consequence, the design of the RMF-P project included a two-day residential professional learning program at the beginning and end of the project for the SNMY specialists and partner representatives. This was supported by online, interactive sessions every 3 to 4 weeks that were open to anyone involved to address issues raised by specialists and to further support the implementation of a targeted teaching approach to multiplicative thinking.

The initial professional learning program was designed to support the SNMY Specialists work with at least two teachers of Year 7–9 mathematics in their respective schools. The aim of this program was to train the Specialists in the use of the SNMY tools, deepen their understanding of multiplicative thinking and proportional reasoning and their connection to other areas of the mathematics curriculum, and explore activities and strategies to support a targeted teaching approach based on the LAF. A two-day professional learning conference was held at the end of the project to document and disseminate project experiences and findings and identify areas for further research.

As one of the objectives of the RMF-P project was to identify what targeted teaching might look like in secondary schools, the specialists and the respective school teams were able to decide exactly how they would use the Zone-based activities to support a targeted teaching approach appropriate to their particular circumstances.

5.1 *Results*

The SNMY assessments were conducted in August and November of 2013 to evaluate the impact of the targeted teaching approach to multiplicative thinking. Matched data sets were obtained from 1732 students from Years 7 to 10 with the majority (59%) from Year 8. Once again, the results were analysed using the Rasch partial credit model (Masters, 1982). Although the results varied considerably between schools, the overall achievement of students across the 28 schools grew above an adjusted[6] effect size of 0.6 indicating a medium influence beyond what might be expected (Hattie, 2012). This can be seen in the shift in the relative proportions in each Zone of the LAF from August to November shown in Figure 5.3.

As indicated above, one of the purposes of the RMF-P project was to explore what targeted teaching might look like in the context of secondary schooling. The following case studies are presented as illustrations of 'what works' in this context.

5.2 *Case Study 1*

Palberton Middle School (not its real name) is located in a growing, outer suburb of a northern Australian city. At the time, the school had 560 Year 7 to 9 students from a diverse range of cultural backgrounds. The school leadership team and the maths staff were keen to improve mathematics learning outcomes and had previously explored a number of commercial intervention programs with limited success. When the opportunity to participate in the RMF-P project was offered they 'jumped at the chance' as they could see this

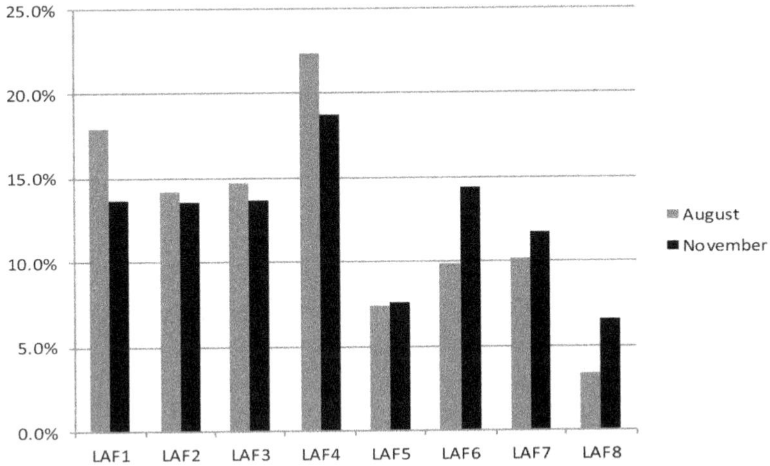

FIGURE 5.3 Proportion of students by LAF zone, RMF-P project (n = 1732)

working well with their commitment to team teaching and visible learning (Hattie, 2012), that is, using data to inform teaching approaches. After attending the initial workshop in Melbourne in July 2013, the specialist teacher in consultation with the school leadership and two other maths staff decided to target four of the Year 8 classes (50% of the cohort) in what remained of the 2013 school year.

The school's purpose-built accommodation facilitated team teaching approaches. Four classrooms were grouped around a central covered space with large sliding doors providing access to the central space from each classroom. Co-teaching arrangements were formalised in 'hub' agreements and the team co-taught two of the four classes while a parallel team of English teachers co-taught the other two classes. This arrangement was reversed at other times in the week so that all four classes had the same team for maths and English. Learning support staff were available on most occasions to support the work of the co-teaching teams. The school timetable provided five 50-minute lessons per week for maths (and English) which included one double lesson.

The RMF team as it became known, administered, marked and moderated SNMY Option 1 for the four Year 8 classes in August 2013 and created profiles for all students. The specialist shared the data with the school leadership team and a key figure in the Department of Education, who were "shocked" to see that 53% of the Year 8 students assessed were in Zones 1 to 3 of the LAF. While the specialist was able to put this into context (the data were consistent with data from similar schools), when the leadership group recognised what this meant, further in-kind resources were made available to support the work of the project.

A decision was made to use the double period in maths each week to implement a targeted teaching approach to multiplicative thinking. These lessons, which came to be referred to as RMF Maths, were structured to include a Do Daily session, an open-ended problem related to the mathematics being considered in the other three lessons, work in Zone groups on targeted teaching activities, and a formal period of reflection. The approximate time spent on each of these components was 10, 40, 40 and 10 minutes respectively. Each member of the team was responsible for two to three Zones. The team met weekly to plan Zone activities, many of which they adapted to be age-appropriate and met again on Saturdays for professional sharing and forward planning. The students were given project books which they decorated in which to record their reflections at the end of each RMF lesson. A template was provided to help structure the reflections. The booklets were collected by the team and returned at the beginning the next lesson. The team reviewed the student reflections and used this to inform their planning. They also provided written feedback for each student on a weekly basis. The team observed considerable changes in the nature and amount of reflective comments provided by the students over the course of the semester – the students looked forward to reading the feedback from the teachers and quickly settled in order to see what was written.

The targeted teaching activities and materials were organised and stored by Zone in the hub area in open shelving that was available to students. This enabled some level of choice if students wanted to move on to another activity or try an activity from another Zone. This was a massive effort but the teachers felt it was worth as one of the first things they noticed were that there were far fewer instances of challenging behaviour to deal with and students were asking if they could do 'RMF maths' all of the time. Another positive outcome was that students were becoming more metacognitive in their responses to problems they were doing in the non-RMF lessons, for instance, they noticed that many of the students started to explain their reasoning without being asked. Although the demands on the teaching staff were high with many additional hours per week spent on preparing and adapting Zone activities, the teachers felt that they had grown as a team and were more knowledgeable about how to deal with student misconceptions.

SNMY Option 2 was administered in November and marked and moderated by the co-teaching team. The data were de-identified, recorded on a spreadsheet and forwarded to the research team for analysis. Data from 70 matched pairs were available for analysis the results of which can be seen in Figure 5.4. The improvement in multiplicative thinking was impressive with an adjusted effect size of 1.18.

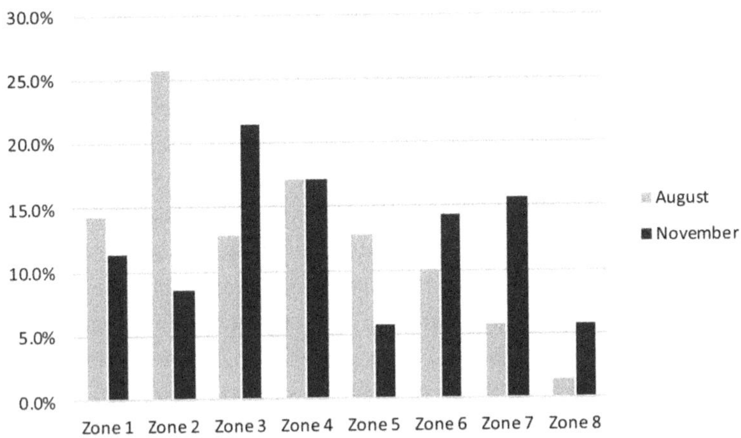

FIGURE 5.4 Proportion of year 8 students by LAF zone, Palberton Middle School (n = 70)

5.3 *Case Study 2*

Plumpton High School (name used with permission) is located in an established western suburb of Sydney. It is a large multi-cultural 7 to 12 secondary school, a key goal of which is to "put students first." The school prioritises English, Maths and Science. It also offers a wide choice of flexible pathways for students to transition into work, further vocational study or university. Plumpton High School came to use the SNMY materials and implement a targeted teaching approach to multiplicative thinking as a result of the school's participation in a follow up study to RMF-P, the *Reframing Mathematical Futures II* project,[7] which was aimed at building an evidenced-based framework for mathematical reasoning in Years 7 to 10 using a similar methodology as that used to develop the LAF. The partners and schools that had participated in RMF-P were offered the opportunity to continue in this project and an invitation was extended to the Departments of Education in New South Wales and Western Australia, both of whom agreed to participate as partners in the follow up project. Six schools from NSW and four from WA were identified as meeting the funding criteria which were the same as for RMF-P. As the 'new' schools were unfamiliar with the SNMY materials and targeted teaching, they were supported to implement the RMF-P project in 2015 and thereafter contribute to and participate in the mathematical reasoning component of the follow up project.

When the opportunity to participate in the follow up project was offered in late 2014, the mathematics results at the school was a concern and number of students electing to pursue the more advanced maths courses in the senior years was declining. A change was needed. As a consequence, the school leadership not only agreed to participate in the project they decided to send an additional teacher to the initial three-day workshop in Melbourne in November 2014 at the school's expense. In addition to setting the scene for the follow-up project, the workshop introduced teachers from the 'new' schools to multiplicative thinking and the SNMY materials. The RMF-P specialists who were continuing in the follow up project were able to share their SNMY results and describe what worked and what did not work in implementing a targeted teaching approach to multiplicative thinking in secondary school contexts. Key strategies that were variously adopted by the RMF-P schools that were most successful included team teaching, dedicated lesson times for targeted teaching and Zone-based activities, locally available resources, team planning time, additional time release, access to professional learning opportunities, and support of school leadership. RMF-P schools implemented these and other strategies to different extents and in different ways appropriate to their circumstances but the teachers from the 'new' schools such as Plumpton were able to draw on this information to plan how they would implement a targeted teaching approach.

On returning to school, the specialist from Plumpton met with the maths faculty and a decision was made to focus on the whole of Year 8 in 2015. Teaching staff felt that the current Year 7 students would most benefit from the intervention and as they were still at school it would make sense to administer SNMY Option 1 in December of 2014 to give teaching staff more time to prepare over the summer break. The school leadership supported the decision to focus on Year 8 in 2015 as this cohort would sit the NAPLAN test in Year 9 in 2016 which would provide an independent evaluation of the intervention.

In 2015, each of the six Year 8 classes had a separate 75 minutes RMF lesson per week. During this time, the students worked in their Zone groupings initially on activities from the project Dropbox and/or ones prepared by the specialist. The specialist and one other of the senior maths teachers, dropped by the classrooms whenever they could to help and prepared resources in their free periods. As time went on and the demand for new, age-appropriate activities increased, the Year 8 teachers also developed and shared Zone-based activities with their colleagues. One of the ways in which this happened was at the Wednesday lunches, where Year 8 staff talked about what they were doing, reflected on progress and developed new ideas. A lesson template was developed and staff would workshop new lessons prior to delivery. Referred to as 'Live in Lessons,' this enabled the team to iron out any potential issues and to make links to regular classroom teaching activities and content.

Project funding was provided to support the implementation of a targeted teaching approach to multiplicative thinking in four classes. However, the school was implementing this approach in 6 classes of 30 students, which meant resources were tight. Priority was given to purchasing concrete materials and a separate area was set up to keep class booklets, resources and activities for easy collection and distribution. From everyone's perspective it was a tough year. There was little buy-in from students and teachers at the outset as working in groups was something new for many. The existing class structure (semi-streamed) helped manage the targeted teaching approach but there was still considerable variation in each classroom. Planning was essential and in retrospect, it was a key factor to the school's success. Over the course of the year, teachers found that they were incorporating many of Zone type activities into the curriculum being taught in the week, placing particular emphasis on the need to explain and justify solution strategies as this had proved to be a major sticking point early on. The team learnt as they went and kept on sharing, adjusting and implementing strategies/activities which worked in other classes. Staff meetings on Mondays were focussed on developing teachers' capacity to share resources and ideas to help the growth of targeted teaching in classrooms.

Gradually, everything became easier, the students were more accustomed to working in groups and appreciated the opportunity to experience success. Student engagement increased and the quality of their responses to school-based assessments improved noticeably. The specialist was able to see considerable changes in her year 8 class in the engagement and results of the students. Teaching staff were more inclined to design reasoning activities for regular classroom teaching and provide time for students to apply what they know in unfamiliar contexts and marking rubrics were slowly incorporated into classroom assessment tasks.

SNMY Option 2 was administered, marked and moderated by Year 8 teachers and the two specialists in September 2015. The results were again de-identified and forwarded to the research team for analysis. The results were impressive and immediately bought buy in from senior management and other maths teachers. Additional teacher release was provided to support the preparation of resources, marking and moderating of assessments, and training of other staff members. In December 2014, 52% of the Year 7 cohort were in Zones 1 to 3 of the LAF. By September 2015, only 30% were in Zones 1 to 3. In 2014, only 16% of the students were in Zones 6 to 8. In 2015, this had risen to 40%. The growth is shown in Figure 5.5 and represents an effect size of over 1.

While not the only measure of success in school mathematics, the Year 9 NAPLAN results for the same cohort in 2016 provide conclusive evidence that targeted teaching makes a difference. Compared to the previous Year 9 who sat the NAPLAN test in 2015, the average scaled growth score for the school went from below all State in 2015 (45.6) to above all State in 2016 (51.1). But perhaps more telling are the respective growth comparisons between 2015 and 2016 of the proportion of students in the less than expected growth category versus the proportion of students in the greater than or equal to expected growth category.

The conclusion that can be drawn from the evidence of the two large-scale studies reported here is that targeted teaching is effective in improving student

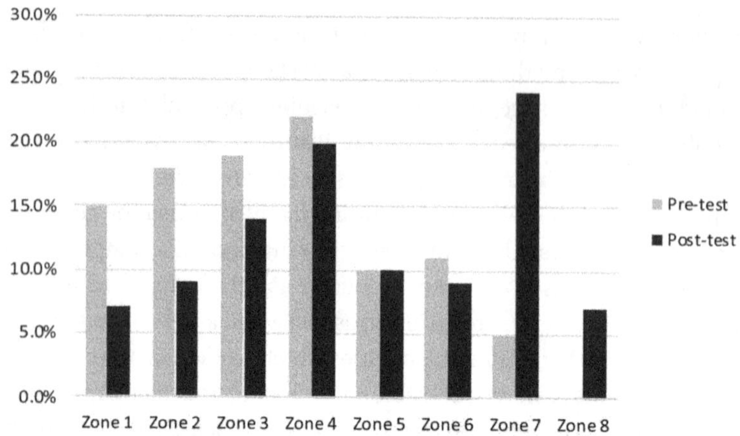

FIGURE 5.5 Proportion of year 8 students by LAF zone, Plumpton High School 2015 ($141 < n < 152$)

TABLE 5.2 Growth comparisons for the 2015 and 2016 year 9 cohorts, Plumpton High School

	Less than expected growth	Greater than or equal to expected growth
2015	48.7%	51.2%
2016	34.5%	65.5%

learning outcomes in mathematics in the middle years. Targeted teaching requires quality assessment tools and evidenced based learning progressions for the big ideas in mathematics, without which student progress will be impacted. However, as the two case studies have shown, this, in itself, is insufficient. Targeted teaching is dynamic, it needs to be implemented in ways that respond to the particular circumstances of the school and the students and teachers within that school. Successful implementations require committed discipline leadership, the support of school leadership more generally and access to quality professional learning.

6 Conclusion

Critical thinking, creativity, communication, and self-direction are all highly desirable outcomes of a STEM education but beyond the primary years they depend upon key underpinning ideas and competencies such as multiplicative thinking, which research over many years has established is responsible for a seven-year range in students' mathematics achievement in the middle years of schooling. Quantification and reasoning about relationships between quantities is essential further study in science, engineering and technology. Participation rates in STEM will not be increased unless and until participation rates in the more advanced mathematics subjects are increased. And those will not be increased while we have declining rates of mathematical literacy among Australian 15-year olds.

The work on learning progressions reported here is aimed at identifying optimal pathways for teaching and learning these key underpinning aspects of school mathematics based on an assessment of what might be regarded as students' taken-as-shared knowledge in Australian mathematics classrooms. A valid criticism of this approach is that it does not necessarily reflect what is possible when students are exposed to high quality mathematics teaching over time (e.g., Boaler, 2008). But the reality is that not all teachers have the knowledge, confidence and local support needed to implement high quality effective practices. Nor do they necessarily have the time and resources to identify each student's particular learning needs in relation to every single aspect of the mathematics curriculum even if this was desirable. The main rationale for working at scale in relation to a small number of really big ideas in mathematics is that this establishes a plausible, probabilistic model for establishing where learners are in their learning journey (Masters, 2013) in relation to the 'big' ideas critical to student's progress in school mathematics (Siemon, Bleckly, & Neal, 2012) and a framework to support teachers progress student learning. In this case the focus is on multiplicative thinking, the aspect of mathematics that is largely responsible for the 7 to 8-year range in student mathematics achievement in the middle years.

The urgency around STEM and STEM education is a global phenomenon as industry and governments grapple with the ever-increasing pace of technological innovation and the digital revolution. STEM education is vital and the view presented here is that it is neither all integrated or all discipline specific. As in all complex domains, the truth lies in a creative and purposeful amalgam of the two – both are needed if we are to create the STEM workforce for the future. An argument has been made for a specific focus on multiplicative thinking as it is key to accessing further mathematics learning but it could also be read as an argument for treating mathematics as a special case within the STEM umbrella because at the end of the day there is no STEM without M.

Acknowledgements

The authors would like to thank Gillian Milne, Sandra Vander Pal, Brian Sharpley and Rosemary Callingham for their contributions to the RMF-P project.

Notes

1. These activities are well developed and have been available for some time as part of the *maths300* suite of exemplary mathematics lessons available from (http://www.maths300.com).
2. The *Middle Years Numeracy Research Project* (MYNRP) 1999–2001 was commissioned by the Victorian Department of Education, Employment and Training (DEET), the Catholic Education Commission of Victoria (CECV) and the Association of Independent Schools of Victoria (AISV) to inform the development of a coordinated and strategic plan for improving the teaching and learning of numeracy in Years 5 to 9.
3. *Scaffolding Numeracy in the Middle Years – An investigation of a new assessment-guided approach to teaching mathematics using authentic assessment* tasks 2003–2006 was an ARC Linkage Research Project awarded to RMIT University in collaboration with the Victorian Department of Education & Training, and the Tasmanian Department of Education.
4. The *Reframing Mathematical Futures Priority Project 2013–2014* was funded by the Australian Mathematics and Science Partnership Programme (AMSPP) scheme sponsored by the Australian Government, Canberra.
5. See https://www.education.gov.au/australian-maths-and-science-partnerships-programme-amspp for further details.
6. Adjusted to one-year as comparisons of effect size in this context relate to what might be expected in one full year of schooling.
7. The *Reframing Mathematical Futures II project 2015–2017* was funded by the Australian Mathematics and Science Partnership Programme (AMSPP) competitive grant scheme sponsored by the Australian Government, Canberra.

References

Anghileri, J. (1999). Issues in teaching multiplication and division. In I. Thompson (Ed.), *Issues in teaching numeracy in primary schools*. Buckingham: Open University Press.

Australian Academy of Science. (2016). *The mathematical sciences in Australia: A vision for 2025*. Canberra: Australian Academy of Science.

Australian Curriculum Assessment & Reporting Authority. (2016). *Australian curriculum: Mathematics*. Retrieved from http://www.acara.edu.au/curriculum/learning-areas-subjects/mathematics

Australian Industry Group. (2015). *Progressing STEM skills in Australia*. Melbourne: Australian Industry Group.

Badley, K. (2009). Resisting curriculum integration: Do good fences make good neighbours? *Issues in Integrative Studies, 27*, 113–137.

Ball, D. (1993). Halves, pieces, and twoths: Constructing representational contexts in teaching fractions. In T. Carpenter, E. Fennema, & T. Romberg (Eds.), *Rational numbers: An integration of research* (pp. 157–196). Hillsdale, NJ: Lawrence Erlbaum Associates.

Battista, M. (1999). The importance of spatial structuring in geometric reasoning. *Teaching Children Mathematics, 6*(3), 170–178.

Baturo, A. (1997). The implication of multiplicative structure for students' understanding of decimal-number numeration. In F. Biddolf & K. Carr (Eds.), *People in mathematics: Proceedings of the 20th conference of the Mathematics Education Research Group of Australasia*. Rotorua: Mathematics Education Research Group of Australasia.

Becker, K., & Park, K. (2011). Effects of integrative approaches among STEM subjects on student learning. *Journal of STEM Education, 12*(5–6), 23–37.

Black, P., & Wiliam, D. (1998). Inside the Black box: Raising standards through classroom assessment. *Phi Delta Kappan, 92*(1), 81–90. Retrieved from http://www.pdkintl.org/kappan/kbla9819.htm

Boaler, J. (2008). Promoting 'relational equity' and high mathematics achievement through an innovative mixed-ability approach. *British Educational Research Journal, 34*(2), 167–194.

Bond, T., & Fox, C. (2001). *Applying the Rasch model: Fundamental measurement in the human sciences*. Mahwah, NJ: Lawrence Erlbaum Associates.

Breed, M. (2011). *Constructing paths to multiplicative thinking* (Unpublished PhD thesis). RMIT University, Melbourne.

Business Council of Australia. (2015). *STEM strategy would help drive innovation*. Retrieved from http://www.bca.com.au/media/stem-strategy-would-help-drive-innovation

Bybee, R. (2013). *The case for STEM education: Challenges and opportunities*. Arlington, VA: NSTA Press.

Callingham, R., & Griffin, P. (2000). *Towards a framework for numeracy assessment*. Adelaide: Mathematics Education Research and Development.

Clark, F., & Kamii, C. (1996). Identification of multiplicative thinking in children in grades 1–5. *Journal for Research in Mathematics Education, 27*(1), 41–51.

Clarke, D. M. (2001). Understanding, assessing and developing young children's mathematical thinking: Research as a powerful tool for professional growth. In J. Bobis, B. Perry, & M. Mitchelmore (Eds.), *Numeracy and beyond: Proceedings of the 24th annual conference of the Mathematics Education Group of Australasia*. Sydney: Mathematics Education Group of Australasia.

Clarke, D. M, Sullivan, P., Cheeseman, J., & Clarke, B. (2000). The early numeracy research project: Developing a framework for describing early numeracy learning. In J. Bana & A. Chapman (Eds.), *Proceedings of the 23rd annual conference of the*

Mathematics Research Group of Australasia. Fremantle: Mathematics Research Group of Australasia.

Cobb, P. (1994). Where is the mind? Constructivist and sociocultural perspectives on mathematics learning. *Educational Researcher, 23*(7), 13–20.

Cohen, J. (1969). *Statistical power analysis for the behavioral sciences*. New York, NY: Academic Press.

Earl, L., & Katz, S. (2006). *Rethinking classroom assessment with purpose in mind: Assessment for learning, assessment as learning, assessment of learning*. Alberta: Western and Northern Canadian Protocol for Collaboration in Education.

Finkel, A. (2017, July). *Measuring up*. Opening address to the 40th Annual Conference of the Mathematics Education Research Group of Australasia, Melbourne. Retrieved from http://www.chiefscientist.gov.au/wp-content/uploads/MERGA-speech.pdf

Furner, J., & Kumar, D. (2007). The mathematics and science integration argument: A stand for teacher education. *Eurasia Journal of Mathematics, Science & Technology Education, 3*(3), 185–189.

Gray, E., & Tall, D. (1994). Duality, ambiguity, and flexibility: A "proceptual" view of simple arithmetic. *Journal for Research in Mathematics Education, 25*(2), 116–140.

Harel, G., & Confrey, J. (1994). *The development of multiplicative reasoning in the learning of mathematics*. New York, NY: Suny Press.

Hattie, J. (2012). *Visible learning for teachers, maximising impact on learning*. Oxford: Routledge.

Kirschner, P., Sweller, J., & Clarke, R. (2006). Why minimal guidance during instruction does not work: An analysis of the failure of constructivist, discovery, problem-based, experiential and inquiry-based teaching. *Educational Psychologist, 4*(2), 75–86.

Lamon, S. (1996). The development of unitising: Its role in children's partitioning strategies. *Journal for Research in Mathematics Education, 27*(2), 170–193.

Lamon, S. (2007). *Teaching fractions and ratios for understanding* (2nd ed.). Mahwah, NJ: Lawrence Erlbaum Associates.

Larson, M. (2017, May). *Math education is STEM education, president's message, National Council of Teachers of Mathematics (NCTM)*. Retrieved from https://www.nctm.org/News-and-Calendar/Messages-from-the-President/Archive/Matt-Larson/Math-Education-Is-STEM-Education!/

Lefkowitz, M. (2014, January 7). Why is math so important? Science, technology and engineering cannot thrive without a solid mathematics education. *Mind Research Institute Blog*. Retrieved from https://blog.mindresearch.org/blog/bid/366158/why-is-math-so-important

Lefkowitz, M. (2015, December 1). The M is STEM: How do we prepare students for jobs in STEM fields. *Mind Research Institute Blog*. Retrieved from http://blog.mindresearch.org/blog/the-m-in-stem

Masters, G. N. (1982). A Rasch model for partial credit scoring. *Psychometrika, 49*, 359–381.

Masters, G. N. (2013). *Reforming educational assessment: Imperatives, principles and challenges*. Melbourne: The Australian Council for Educational Research (ACER). Retrieved from http://research.acer.edu.au/aer/12

Misailidou, C., & Williams, J. (2003). Diagnostic assessment of children's proportional reasoning. *Journal of Mathematical Behaviour, 22*, 335–368.

Mulligan, J., & Mitchelmore, M. (1997). Young children's intuitive models of multiplication and division. *Journal for Research in Mathematics Education, 28*, 309–331.

OECD. (2013). *International benchmarking for school improvement: The OECD test for schools*. Paris: OECD.

Office of the Chief Scientist. (2012). *Mathematics, engineering and science in the national interest*. Canberra: Australian Government.

Office of the Chief Scientist. (2013). *Science, technology, engineering and mathematics in the national interest: A strategic approach*. Canberra: Australian Government.

Office of the Chief Scientist. (2014). *Science, technology, engineering and mathematics: Australia's future*. Canberra: Australian Government.

PricewaterhouseCoopers. (2015). *A smart move: Future proofing Australia's workforce by growing skills in Science, Mathematics, Engineering and Maths (STEM)*. Retrieved from http://www.pwc.com/gx/en/industries/government-public-services/public-sector-research-centre/australia/smart-move.html

Reilly, Y., & Parsons, J. (2011). Delivering differentiation in the fully inclusive middle years' classroom. In J. Carroll (Ed.), *Maths is multi-dimensional, Proceedings of the 48th annual conference of the mathematical association of Victoria* (pp. 25–32). Melbourne: Mathematical Association of Victoria.

Rosicka, C. (2016). *From concept to classroom: Translating STEM education research into classroom practice*. Melbourne: The Australian Council for Educational Research (ACER).

Sanders, M. (2009). STEM, STEM education, STEM mania. *Technology Teacher, 68*(4), 20–26.

Siemon, D. (2013). Launching mathematical futures: The role of multiplicative thinking. In S. Herbert, J. Tillyer, & T Spencer (Eds.), *Mathematics: Launching futures, Proceedings of the 24th biennial conference of the Australian Association of Mathematics Teachers* (pp. 36–52). Adelaide: Australian Association of Mathematics Teachers.

Siemon, D. (2016). *Addressing the STEM challenge through targeted teaching: What's the evidence?* Proceedings of the 2016 ACER Research Conference, Melbourne.

Siemon, D., Bleckly, J., & Neal, D. (2012). Working with the big ideas in number and the Australian curriculum: Mathematics. In B. Atweh, M. Goos, R. Jorgensen, & D. Siemon (Eds.), *Engaging the Australian national curriculum: Mathematics – Perspectives from the field* (pp. 19–45). Melbourne: Mathematics Education Research Group of Australasia.

Siemon, D., & Breed, M. (2006, November). *Assessing multiplicative thinking using rich assessment tasks*. Paper presented to the Annual Conference of the

Australian Association for Research in Education, Adelaide. Retrieved from http://www.aare.edu.au/data/publications/2006/sie06375.pdf

Siemon, D., Breed, M., Dole, S., Izard, J., & Virgona, J. (2006). *Scaffolding numeracy in the middle years: Project findings, materials and resources, final report.* Retrieved from http://www.education.vic.gov.au/school/teachers/teachingresources/discipline/maths/assessment/Pages/scaffoldnum.aspx

Siemon, D., & Virgona, J. (2001, December). *Roadmaps to numeracy: Reflections on the middle years numeracy research project.* Proceedings of the Annual Conference of the Australian Association for Research in Education, Fremantle. Retrieved from https://www.aare.edu.au/data/publications/2001/sie01654.pdf

Simon, M. (1995). Reconstructing mathematics pedagogy from a constructivist perspective. *Journal for Research in Mathematics Education, 26*(2), 114–145.

Sullivan, P. (2011). *Teaching mathematics: Using research-informed strategies* (Australian Education Review No. 59). Melbourne: The Australian Council for Educational Research (ACER).

Sullivan, P., Clarke, D., Cheeseman, C., & Mulligan, J. (2001). Moving beyond physical models in multiplicative reasoning. In M. van den Heuvel-Panhuizen (Ed.), *Proceedings of the 25th conference of the International Group for the Psychology of Mathematics Education* (Vol. 4, pp. 233–241). Utrecht: International Group for the Psychology of Mathematics Education.

Thomson, S., De Bortoli, L., & Buckley, S. (2013). *PISA 2012: How Australia measures up.* Melbourne: The Australian Council for Educational Research (ACER).

Thomson, S., De Bortoli, L., & Underwood, C. (2017). *PISA 2015: Reporting Australia's results.* Melbourne: The Australian Council for Educational Research (ACER).

Thomson, S., Wernet, N., O'Grady, E., & Rodrigues, S. (2017). *TIMSS 2015: Reporting Australia's results.* Melbourne: The Australian Council for Educational Research (ACER).

Venville, G., Rennie, L., & Wallace, J. (2009). Disciplinary versus integrated curriculum: The challenge for school science. *The New Critic, 19*, 1–9.

Vergnaud, G. (1988). Multiplicative structures. In J. Hiebert & M. Behr (Eds.), *Number concepts and operations in the middle grades* (pp. 141–161). Reston, VA: National Council of Teachers of Mathematics.

Vergnaud, G. (1994). Multiplicative conceptual field: What and why? In G. Harel & J. Confrey (Eds.), *The development of multiplicative reasoning in the learning of mathematics* (pp. 41–60). Albany, NY: State University of New York.

Wai, J., Lubinski, D., & Benbow, C. (2009). Spatial ability for STEM domains: Aligning over 50 years of cumulative psychological knowledge solidifies its importance. *Journal of Educational Psychology, 101*(4), 817–835.

CHAPTER 6

Possibilities and Potential with Young Learners: Making a Case for STEAM Education

Andrew Gilbert and Lisa Borgerding

Abstract

This case study delves into a five-day STEAM camp at a Reggio Emilia inspired pre-school setting where children explored STEM content that included a strong Arts component. The results suggest that integrated STEAM activities helped young children construct understandings for the properties of air and facilitated engagement in argumentation surrounding those concepts. Young children demonstrated rich possibilities and potential for learning across STEAM and this project serves as a reminder that our youngest learners are capable of engagement in STEM particularly when explored using the Arts for design, testing and communication of their burgeoning ideas.

Keywords

Early childhood – STEAM – Reggio Emilia – air – arts

1 Introduction

Many education professionals intuitively understand the value of STEM in early childhood education (ECE) contexts, but typically face obstacles regarding resources, time and teacher content confidence in regards to creating viable STEM approaches with young children (Linder, Emerson, Heffron, Shevlin, & Vest, 2016). This chapter documents a week-long STEAM program carried out in a Reggio Emilia (RE) inspired pre-school in a North American context. The goals for this chapter are to describe key areas of importance in regard to STEM in ECE contexts and present arguments for the consideration of STEAM with pre-school age children. This will contextualize the work described within the chapter and articulate the potential to engage children in age-appropriate integrated approaches to content. The chapter includes detailed descriptions for pedagogical approaches utilised by authors and teaching staff, as well as

 | DOI:10.1163/9789004391413_007

documented observations of children during these STEAM encounters. These approaches were carried out in a North American university-based laboratory preschool setting as a means to articulate that young children are well-positioned to engage in STEAM learning opportunities.

2 Intersection of STEM and Early Childhood Education Principles

2.1 *Appropriateness of STEM in Early Childhood*

Cobb (1977) reminds us of the genius of childhood and the imaginative brilliance utilised by children during their interactions with the natural world and asserts that what is essential in the education of young children is their growing intellectual awareness that demonstrates an, "evolving ability to learn, think, and create meaning in his perceived world, in contrast to the ability to memorise and record other people's interpretations of the world" (p. 18). We are reminded of the power of children and their ability to think through complex situations. We argue that engaging young children in meaningful STEM-related challenges is one key feature to bringing children to their fullest future potential. The authors and the preschool, within which this work was embedded, shared Cobb's vision for both the ability and potential of children.

Namely, the university-based preschool clearly embodied the spirit and philosophical positioning of Reggio Emilia (RE) inspired experiential approaches to engaging children in learning. Gandini (1993) articulated a clear vision of an RE philosophy where,

> children have preparedness, potential, curiosity, and interest in constructing their learning, in engaging in social interaction, and in negotiating with everything the environment brings to them. Teachers are deeply aware of children's potentials and construct all their work and the environment of the children's experience to respond appropriately. (p. 5)

This respect for children's abilities to build knowledge and understanding for their surroundings is an essential and enduring foundation for RE. Preschools adhering to an RE philosophy embody these notions, but generally speaking this experiential approach is represented by two main pedagogical constructs: (1) designing an initial experience that is rich in resources and provides freedom for children to enact their thoughts and ideas; (2) observing and documenting children's ideas to build further learning progressions (Hong, Shaffer, & Han, 2017). Close observation and documenting children's engagement with content are the hallmarks of RE's emergent inquiry process. There is also a clearly held reverence and respect for the ability of the young child that permeates pedagogical intentions in the classroom. This deeply held belief "builds on the premise that each child has the desire to connect with others, to engage in learning, and to enter into a relationship with their environment" (Dodd-Nufrio, 2011, p. 236). Teachers in RE classrooms work to achieve these pedagogic principles by facilitating children connecting with one another through purposeful experiences and scaffolding language that fosters engagement with the natural world. These typically entail an experiential session with an expectation for children to generate ideas and openly test their hypotheses about those phenomena. Based on these experiences, children build their final ideas and solutions and express those ideas through explanation and representation of their experience and resulting solutions to the original problem or challenge, which is an essential fit for engaging young children in inquiry (Inan, Trundle, & Kantor, 2010).

One of the key reasons that preschool contexts, steeped in RE, are an important context for STEM engagement is that teachers in these contexts have a predisposition for integrated thinking and a willingness to trust in the thinking ability of young children. Linder, Emerson, Heffron, Shevlin, and Vest (2016) articulated that ECE teachers' positive dispositions toward integration makes them uniquely prepared to meet the challenge of engaging in STEM. Thus, we built this work on the notion that exploration and inquiry are a natural state of learning for young children, where intentional planning within STEM can build a deep understanding for the role of evidence to make claims about the natural world (Bosse, Jacobs, & Anderson, 2009). However, despite their positive disposition there still exists a need for ECE teachers to develop their abilities for implementing inquiry and engineering design (Tuttle et al., 2016).

2.2 *Operationalising STEM in Early Childhood: Making a Case for STEAM*

Vasquez, Sneider, and Comer (2013) posited that the hallmark of engineering practices is that there are multiple possible solutions to design challenges. RE classrooms are particularly aligned with this openness to multiple solutions and a multiplicity of competing ideas working toward solutions and RE teachers regularly incorporate these approaches in their everyday practices. In terms of pedagogy, Moomaw (2013, p. 5) identified four principles for building and implementing integrated STEM practices with young children: (1) intentional teaching; (2) teaching for understanding; (3) encouraging inquiry and (4) providing real-world contexts. There also exists considerable overlap between Moomaw's vision for STEM best practices and the tenets of RE. For instance, in terms of intentional teaching, teachers must be keenly aware of the content goals and skills that children will engage with during the designed learning opportunity. This equates with RE's demand that teachers need to pay careful attention to children and their ideas in order to build scaffolded activities as children's knowledge and questions develop. In addition, principles 2–4 also clearly align with RE in terms of engaging children in emergent inquiry and connecting those directly to the lives and experiences of children.

There exist special challenges of bringing STEM practice into ECE contexts in terms of literacy and ability to express ideas in writing. Thompson (1995) demonstrated how sketching and artistic expression were an essential and important aspect of children's developing literacy as well as a means to document thinking. The arts rightfully occupy a major portion of any ECE classroom, but it also offers special connection to STEM content. In this regard, the inclusion of the arts provides a mechanism to interest current ECE teachers in STEM related content, "Since the arts are a natural part of early childhood education, adding this element may help more teachers find ways to work STEM concepts into the curriculum" (Sharapan, 2012, p. 36). The Arts then become a way to span children's ability to express their ideas and offers a conduit to STEM learning and problem-solving approaches. "The inclusion of the Arts in ECE problem-solving revolves around ... Drawing, painting, modeling, and construction are all used to deepen the children's understanding of the topic and allow them to represent their understandings in concrete ways" (Greibling, 2010, p. 6). In addition, the inclusion of the Arts also has definitive connections to content and process involved in the STEM fields. For instance, Gess (2017) argued that artistic artifacts are constructed through a design process that, "is inherent in both engineering and artistic endeavors. It stands to reason, therefore, that either engineering or art may be used as a context in which meaningful learning may occur" (p. 40). Consequently, we utilise the term "STEAM" within this chapter to best articulate the centrality of the arts as a means for children to express and engage with their thoughts regarding content in developmentally appropriate ways.

The overlapping theoretical and pedagogical spaces of RE and STEAM provided the foundation for a weeklong STEAM camp experience with children between the ages of four to six years old at an RE inspired research-based pre-school. Tippet and Milford (2017) argued that there is precious little research carried out in ECE in regard to STEM learning in general, but particularly across North American contexts. Therefore, our goals are to provide some insights into North American pre-school contexts to articulate possibilities and potential of STEAM with young children. This STEAM experience was intentionally focused on a single content theme, namely the conception of *air*. We chose to focus our research effort on a single concept and build the associated STEAM activities within that theme to delve deeply into the possibilities and impacts on learning and teaching in ECE. The goal is to demonstrate both the potential of STEAM in ECE for promoting children's learning and engagement and that these approaches are not out of reach for teachers of ECE.

3 Descriptions of the STEAM Camp: Context and Approach

In order to better understand the children's experiences and engagement with STEAM activities, associated with air, we utilised a case study approach (Stake, 2000). To this end, we collected ethnographic data that included detailed observations, photos, note taking and reflective memos by both the researchers and classrooms teachers. This also included two pre-service teacher assistants assigned to the classroom. These data sets were then subjected to multiple readings to build preliminary categories (Miles & Huberman, 1994). These initial categories where then subjected to constant comparison and analysis to develop emergent themes (Strauss & Corbin, 1998), which are represented in the findings of this chapter.

The STEAM (Science, Technology, Engineering, Art, and Mathematics) camp was a 10-week summer program for children aged 3 through 6. The STEAM camp took place at a child development center associated with a large Midwestern university in North America, and most of the children attending the camp were children of faculty, staff, and students from this university. Children were registered for the camps week by week, although most participants participated in each of the STEAM camp sessions. This chapter focuses on the fifth week of that program where children focused solely on the conception of *air*.

The following description was provided to parents as a means of promoting this week of the program:

> *The Wonders of Air*
>
> This interactive experience will investigate the presence, nature and properties of an essential, but often little understood aspect of our everyday lives: Air. The sessions will revolve around children's wonders regarding air and help them develop methods to investigate those wonders.

Children were divided into two sections, based on their developmental need for afternoon naps. Thus, the younger class included the three-year-olds and four-year-olds who still napped; the older class included the non-napping four-year-olds, five-year-olds, and six-year-olds. For the purposes of this chapter, we will focus solely on the older classroom children which was comprised of 23 children (9 boys, 14 girls). In terms of ages, there were six four-year-olds, 13 five-year-olds, and four six-year-olds. The authors were instructor-researchers for this camp as we planned, implemented, and gathered data throughout this particular camp week.

3.1 *Wonders of Air Camp (Older Class)*

The primary topics for the camp are included in Table 6.1. Each day started off with a Morning Meeting during which the day's topics were introduced through children's books and demonstrations. Children were prompted to wonder about the phenomena, and their wonders were recorded on the Wonder Wall (Cooper & Gilbert, 2016). Following the morning meeting, children rotated freely through various centers built around the day's topics.

TABLE 6.1 Activities, schedule and approach for wonders of air camp

Day	Main topics	Morning meeting	Center choices	Wrap-up
1	Wondering about air What and where is air? Playing with air	– Introduce Wonder Wall – Read "I Wonder" book to generate wonders about air – Paper Bag Assessment: Is the paper bag empty?	– Using air to move things: straws and ping pong balls – Air inside: balloons/beach balls – Bed sheet air: making air move – Dancing ribbons: movement in air – Bubble station: air inside & moving air	– What did we notice about air?
2	Feeling air Moving air Air around us	– Wonder Wall – Vote with feet: Does air move? – Balloon & fan: predictions about how the balloon will move with the fan on and off – Helicopter WHO: What does this make us wonder? How can we test our ideas?	– Bed sheet challenges: – Making and testing helicopters – Making heat spirals – Ping pong balls up and down ramps	– Parachute: describing air movement – Air scavenger hunt: finding moving air

(*cont.*)

TABLE 6.1 Activities, schedule and approach for wonders of air camp (*cont.*)

Day	Main topics	Morning meeting	Center choices	Wrap-up
3	Wind Testing Moving Air	– Debrief helicopter movement – Read "It Blew in the Wind" – Wondering about fan blades and pinwheels (wonder wall) – How can we build something to test if air is moving?	– Engineering task: Detecting moving air – Wind: helicopter movement – Pinwheel construction and testing (fan at different speeds) – Windsock construction and testing (fan at different speeds)	– How did our tests go? – Read "A Windy Day" and observations of moving air outside
4	Can air push?	– Wonder: Can air push? – Read "I Face the Wind" with sandwich bag investigations – Air pressure demos: water tub with bottles, upside down cup with notecard, keeping paper towel dry, bottle & the bag, Cartesian diver	– Free play with air pressure demos – Making parachutes for toys – Face races – Walk with paper	– Read "Talk about Air" and debrief "pushy" air – Demo: Ping pong ball and hair dryer – Sticker assessment
5	Hearing air Sound	– Wonder: can we hear air? – Read "Listen to the World" – Music instrument presenter/ participation	– Making drums – Tuning fork tests – Rubber bands (vibrating air) – Kazoo construction with straws	– Orchestra demo – Fog cannon: what is inside the box?

When these centers involved engineering tasks or new activities, one of the teachers or teaching assistants staffed that particular table. At the end of each day, the children returned to a whole-group setting for a wrap-up session that debriefed the day's activities. We enacted these approaches with a team of educators all of whom subscribed to a Reggio approach and were led in terms of STEM content by the authors of this piece. The team consisted of a classroom ECE teacher well versed in RE, graduate student pre-service teacher, and the authors. This team worked in concert to plan, coordinate and document student learning throughout the camp. What follows are illustrative examples from the second author's work with the older class and demonstrates how the team and children engaged in these activities and how those activated STEAM learning with the concept of air at the core of the investigations.

4 Actualisation of STEAM with Young Learners: Two Illustrations

4.1 *Engineering Task – Testing Whether or Not Air Moves?*

On the second day of camp, the children observed air movement in several capacities. During the third day's session, we discussed the phenomenon of "wind" as moving air. They wondered many things about moving air and wind. They observed that air "moves when you walk, run, and jump" (four-year-old). When probed specifically about wind, one child asserted that wind "brings rain" (four-year-old). I asked where wind comes from, and children offered that it "comes from clouds" (five-year-old) and that "when a storm is coming, it pushes the wind" (six-year-old). One five-year-old boy realised that wind – moving air – comes "from everywhere – our bodies, inside, and outside."

I asked children what they wondered about air, and their emergent questions generally focused on "how does the wind move?" Many of the children's wonders centered around the visibility of wind. One six-year-old girl asked, "Why can't we see the wind?" Other children conjectured that "wind is clear" (five-year-old) and that "wind is invisible" (five-year-old). Another child (five-year-old) insisted that "wind is so small you can't see it" and pleaded with me to bring in a microscope so we could see wind. This issue of making wind visible led to the development of a challenge and investigation process designed to scaffold student understanding concerning the properties of wind. I asked, "if we can't see air, how can we tell if it is moving?" The children were not sure, and I reframed this question as a challenge: I asked the children to consider how they could build something to test whether or not air was moving.

We provided children with many materials including coffee filters, balloons, straws, plastic and paper sacks, cardboard from recycling, sheets of paper, and pipe cleaners. For assembly, we provided children with tape and glue. Most children made an initial construction, tested it and watched other children's prototypes, and then reconstructed and tested another prototype. To test their constructions, two or three children at a time came over to a fan set at a slow setting. The children placed their construction in front of the fan, usually two or three times. As they were testing their constructions, we recorded what children were saying about their designs and the movement of air. Some of the children's initial designs did not move except for falling when tested in the moving air of the fan. When a five-year-old's construction paper-cardboard structure fell to the ground, he explained how his design was like a "rock – it's too heavy."

Several children's initial designs did not move with the air because they were too heavy. But, when one child (six-year-old) glued a coffee filter onto construction paper and taped a plastic bag to it, she described how "it's a parachute" and "it blows away." After other children saw this child's successful prototype, they began attaching plastic bags to their constructions as well. Soon thereafter, a five-year-old described her construction as "mov[ing] fast" and saying her plastic bag construction was a "flying net." Some children even used some initial "push/pull" language to describe the successful movement of their constructions. For example, another five-year-old explained how his "bag takes the air. It can take the air, so it can *push* it hard."

As a wrap-up discussion, I asked the children if their constructions were able to show whether or not the air was moving. Most children had at least one successful construction, and they affirmed this in the meeting. They concluded that some materials (coffee filters and plastic bags) were more useful than others (construction paper and straws) for building a device to detect the movement of air. In terms of STEAM in ECE, we argue that the engineering task of designing an apparatus that can prove that "air moves" provided pathways into integrated thinking. For instance, undertaking a design-based task (engineering) involved exploring the concept and properties of air itself (science), using multiple materials and tools during construction of prototypes (technology), measurement/probability and data collection (mathematics) and all of these activities are carried out through the contexts of the arts (Greibling, 2010; Gess, 2017). Consequently, the integrated nature of STEAM provided powerful experiences for young children to build meaning and scaffold some beginning understandings for the conception of air.

4.2 *Argumentation – Does Air Push?*

Children engaged in thoughtful STEAM argumentation as they negotiated whether or not and in what contexts air could *push* on objects. Throughout this conversation, children made claims (sometimes disputing each other) and supported these claims with evidence. When I presented them with some discrepant events that refuted their claims, they revised their explanations in light of this new evidence.

The conversation began when I asked the children, "Does air push?" Several children nodded and said "yes." A five-year-old girl agreed and provided the evidence that air pushes down "when you're trying to jump." A six-year-old girl said that air "pushes you when you are running – so fast in the air." I asked children, "Does air push when you're still?" To this, many children vocalised "no." Then a six-year-old girl said, "yes." A five-year-old boy said, "if you are still outside" then the air still pushes on you. A five-year-old conceded and said, "air pushes you down," and a four-year-old agreed with her that "air pushes you down."

I asked if air only pushes people when they are outside, and many children indicated that air only pushes on people when they are *moving* outside. A five-year-old disagreed and said, "you feel it standing still outside." Some of the children agreed, while many other insisted that air only pushes on people when they are moving outside.

We read "A Windy Day," and I invited the children to try to catch air in the plastic bags. Some children tried to blow air into their bags, but most of these attempts failed as air escaped out the wide mouth of the bags. Other children just tried closing the bags, and some were able to trap air inside. Eventually, some children swung their plastic bags around and closed them quickly, trapping air inside. As a couple of children were successful with this process, other children replicated the motions until everyone had some air trapped in their bags.

Once everyone had some air, I asked the children, "if we hold onto the plastic bags in one hand and push on them with our other hand, what will happen." Children predicted that the bags would move or pop as their balloons had done on a previous day. I asked, "If we push on the bags, will they push back?" The children thought about it, and a couple of children predicted that the plastic bags would push back. We tried several times and concluded that the bags pushed back. I asked the children, "What is pushing back at us?" Many children responded that the baggies pushed back. I asked them if the empty-no-air baggies would push back, and the students thought about this, tried it, and determined that the empty bags do not push back. So, I asked the children to again think about what was pushing back in the bags full of air. One six-year-old girl said "air," and several other agreed. The presence of air was, after all, the only difference between the empty and full bags. I also asked, are we inside or outside, and children replied that obviously we were inside. So, I then asked them, "so do you think that air only pushes on us when we are outside?"

We next engaged in a series of demonstrations. In the first demonstration, I had a large coffee can with a Ziploc plastic bag sealed onto the top. The bag was inflated with air. Once the children agreed that there was air in the can/bag construction, I asked the children, "Does the air make a push?" Most children were unsure until they pushed against the bag, and they all agreed that the air pushes back. For the second can demonstration, I had the same coffee can/bag set up, but the bag was sealed with no air on the inside. When the children pulled on the bag, the low pressure in the bag made it very hard to move. When I asked the children if there was air pushing in this example, they were not sure.

For the next set of demonstrations, I held up two sheets of paper – one horizontally and one vertically – and asked children to predict which one will fall to the ground first. Children were mixed on their predictions, and they all observed that the vertical sheet fell the fastest. I asked them if air was pushing on the horizontal paper to slow it down? Some children agreed. When I asked what direction the air was pushing, some children verbalised that the air was pushing up. I then held out a picture of a cartoon bear wearing an inflated parachute. I asked children what was pictured. A five-year-old boy immediately said, "parachute" and described several army movie examples of parachutes. He explained that people wear parachutes, so they can land safely, and we discussed whether or not the person would fall more quickly or slowly while wearing a parachute. I had a homemade parachute made of a plastic sack attached to a small cardboard doll. Asked the children to watch me drop it and make observations. I asked them "where is the air?" and "is the air pushing?" Two five-year-old boys and a six-year-old girl immediately said that the air was pushing up. This introduction led into a construction activity where children made and tested parachutes. Again, the integrated nature of STEAM came via the Arts (Greibling, 2010; Gess, 2017). As such, children jumped into building/designing their own parachutes and testing differing materials, looking for patterns and predictability of their new creations all the while building new understandings for the properties of air itself.

We provided these illustrations to position the reader to better understand both the pedagogical decisions of the teacher and the resulting STEAM engagement with the children involved. The questions that arose and how children pursued answers to those questions and possibilities. Conceptions of air are often misconceived by people across a broad array of ages (Rollnick & Rutherford, 1993) and as such provided an interesting topic for us to tackle with young children in regard to multiple directions available to the children to test their thinking. The Arts provided a powerful context to engage children in science, technology, engineering and mathematics. These illustrations are emblematic of the processes that took place over the entire week on which we frame the following implications for researchers interested in approaches STEM/STEAM with young learners as well as ECE teachers who may wish to incorporate integrated STEAM approaches.

5 Special Considerations for Integrated STEAM in Early Childhood

5.1 *Centrality of Reggio Emilia Principles in Integrated STEAM*

The following discussion highlights key considerations for researchers and educators thinking of incorporating STEM/STEAM approaches in ECE contexts. We provided these approaches with ECE classrooms as a means to provide insights into the depth and complexity of young children's thinking when engaging with STEAM. These learning scenarios were carefully planned and scaffolded as expected in a RE approach (Hong, Shaffer, & Han, 2017) and worked to introduce and support key notions regarding the central role that STEAM can play with our young learners. These processes were seamlessly integrated as they each draw from active thinking and engagement with phenomena, we do not pretend that context did not matter. The Children Centre's deep commitment to RE influenced pedagogy directly supported the goals and processes of STEAM-related inquiry. This includes the social constructivist ideals, where children are used to sharing ideas publically and talking through solutions to problems. This public form of argumentation is an essential aspect of scientific thinking and a central tenet to the process of building and testing hypotheses in the sciences.

RE is also clearly positioned to enact design principles and processes involved in engineering. Tuttle et al. (2016) argued that there exists a need to further develop ECE teachers' awareness and abilities in engineering design; however, we argue that ECE teachers who are committed to an RE philosophy already exhibit many of these design practices in their classrooms. The key for RE teachers is that they utilise the Arts for both expression of explanations and the manipulation of materials as a problem-solving tool. This RE foundation creates an effective context to superimpose integrated STEAM content and in our case the teaching team worked together to craft the camp experience. ECE teachers provided feedback to plans and activities about children's development and capability levels and the researchers provided content background expertise. The researchers led the teaching with on-going real time support from classroom teachers to help collect data, observe student ideas, and help researcher teachers to scaffold content as well as guide pedagogical decisions during the STEAM camp. This collaborative effort was essential toward ensuring children's engagement with the content in ways that resonated with children's abilities and experience. For us, it is important to build ECE teacher's content background, but university researchers also must pay careful attention to the particular challenges and possibilities with young children, and ECE teachers are essential in translating this boundary. In addition, it is essential for researchers in these contexts to respect, acknowledge and learn from the ECE classroom teachers. The classroom teachers provided important knowledge about individual learners' interests, routines, etc. that helped us meet our learners where they were prior to the camp.

There is much promise in these integrated approaches with young children as we nurture their ways of thinking, acting and being in regard to sense making and investigation of the natural world. This foundation is essential to future STEM learning.

5.2 *Pedagogical Considerations*

This work has led to findings that those wishing to undertake STEM/STEAM related research should consider when working in ECE contexts. These will be apparent to many who work with young children, but as STEM researchers attempting to work with young children and/or teaming with ECE educators we provide the following insights to help manage working in a context that most in the STEM field may have limited experience.

1. *Attention to each child:* This includes listening, questioning and engaging with every child's individual ideas. In general, this includes having children verbalise their ideas and utilise materials to represent their thinking. The key here is to scaffold children's experience and connect to STEM content while simultaneously keeping the child's ideas in the foreground. The goal is to support the development of content related thinking, but also build up the child as inquisitive thinker that will serve their future STEAM related identity formation by engaging the child's wonders, questions and burgeoning hypotheses (Gilbert & Byers, 2017).
2. *Need for a staging area:* This area serves as a set up area for each day's events, but also provides quick access for immediate additions and changes to the activities. This is a practical consideration for meeting the needs of multiple children and being responsive to students' ideas. In order to scaffold children's thinking, teachers must be ready and able to adapt and construct experiences that help strengthen and deepen STEM thinking. During this camp, this staging area permitted us to preserve materials from one day to the next in order to revisit particular activities iteratively. This requires access to a myriad of readily available resources particularly in terms of Arts integration and Engineering design. The ready access to support materials aids in designing discrepant events and providing materials for students to test their ideas.
3. *Need for prepared prompts and backup activities:* In order to be prepared to move at the speed of children's thinking and to elicit a myriad of student

ideas, it is essential to have prepared questions and activities to support the content. Often these may never be used but in many cases, it offers key questions and tasks that are available to the teacher on a moment's notice. This level of preparation is essential when trying to manage content goals while simultaneously creating engaging tasks that are responsive to children's ideas.

4. *Assessment challenges:* STEM researchers, who are not also ECE educators, will need to rethink how they consider assessment approaches with young children. These assessments should be highly individualised and document what children are saying and how they are interacting with materials as they develop their ideas. This form of careful documentation of children's thinking, inherent in RE, is a powerful form of assessment (Rinaldi, 2006). This gives important insight into children's thinking, but the difficulty is that it takes careful attention and a labour-intensive approach for making sense of each child's engagement. This is where additional people who are able to document children's responses and questions are an invaluable addition to the teaching context. These insights are key in shaping the direction of future practice with the children.
5. *Reflection and dialog:* A key component of our work with the camp children leans heavily on RE's primacy of reflection and dialog (Seidel, 2001). This is a serious commitment that must also be met by those wishing to engage young children in meaningful STEM/STEAM approaches. The authors engaged in dialog and reflection before, during and after each teaching session. This provided important conversations about what children were articulating, questions that emerged from each day's feedback from students, and how these experiences and questions should influence our future approaches.

6 Conclusions

The main goals of this piece were to articulate the possibilities and potential for integrated STEAM approaches with young children. The two detailed illustrations regarding *Does air move?* and *Argumentation* provided insights for the depth and complexity of children's thinking and their ability to engage in testing their ideas. We argue there exists incredible potential to bring exciting and impactful STEAM practices to young children particularly if those ECE contexts are also steeped in RE philosophy and pedagogy. The overlap between RE's insistence on observing the natural world and encouraging children to build their own solutions (with scaffolding and support from adults and peers) to meaningful tasks becomes a perfect companion to both the Nature of Science and Engineering Design. For instance, we would often have children *wonder*, *hypothesise*, and *observe* natural phenomena or a discrepant event, whet we termed 'WHO structures.' The outcomes from this thinking structure than became the foundation on which children would engage in engineering practices, using the Arts, as they designed solutions and tasks concerning the phenomena. Cyclical processes, such as these, provided opportunities for children to grow and evolve over time, where the process did not end with one simple answer rather it opened up new questions and possibilities to consider (Whitin & Whitin, 1997). Helping our youngest learners internalise these mindsets that comprise the work of scientists, mathematicians and engineers is an essential step in bringing STEM learning to a more diverse and innovative future. This also provides important avenues for future research regarding the associated impacts that engagement with STEAM can have on children's long-term association with STEM thinking and learning as they age. The children involved in this work demonstrated rich possibilities and potential for STEAM learning in design and the pursuit of answers to problems as well as access to sophisticated conceptual understandings even in our youngest stages in life. Consequently, we argue that ECE contexts represent important potential for STEM as an emerging field of inquiry where the Arts must play a vital role in the learning journeys of young children.

References

Bosse, S., Jacobs, G., & Anderson, T. (2009). Science in the early years. *Young Children, 63*(4), 10–15.

Cobb, E. (1977). *The ecology of imagination in childhood.* New York, NY: Columbia University Press.

Cooper, G., & Gilbert, A. (2016). Using moments of wonder in science with pre-service teachers. *Asia-Pacific Forum on Science Learning and Teaching, 17*(2), 1–27.

Dodd-Nufrio, A. (2011). Reggio Emilia, Maria Montessori, and John Dewey: Dispelling teachers' misconceptions and understanding theoretical foundations. *Early Childhood Education Journal, 39*, 235–237.

Gandini, L. (1993). Fundamentals of the Reggio Emilia approach to early childhood education. *Young Children, 49*(1), 4–8.

Gess, A. (2017). STEAM education: Separating fact from fiction. *Technology and Engineering Teacher, 77*(3), 39–41.

Gilbert, A., & Byers, C. (2017). Wonder as a tool to engage preservice elementary teachers in science learning and teaching. *Science Education, 101*(6), 907–928.

Greibling, S. (2010). Discoveries from a Reggio-inspired classroom: Meeting developmental needs through the visual arts. *Art Education, 64*(2), 6–11.

Hong, S., Shaffer, L., & Han, J. (2017). Reggio Emilia inspired learning groups: Relationships, communication, cognition and play. *Early Childhood Education Journal, 45*(5), 629–639.

Inan, H., Trundle, C., & Kantor, R. (2010). Understanding natural sciences education in a Reggio Emilia-inspired preschool. *Journal of Research in Science Teaching, 47*, 1186–1208.

Linder, S., Emerson, A., Heffron, B., Shevlin, E., & Vest, A. (2016). STEM use in early childhood education: Viewpoints form the field. *Young Children, 71*(3), 87–91.

Miles, M., & Huberman, A. (1994). *Qualitative data analysis: An expanded sourcebook* (2nd ed.). Thousand Oaks, CA: Sage Publications.

Moomaw, S. (2013). *Teaching STEM in the early years: Activities for integrating Science, Technology, Engineering and Mathematics*. St. Paul, MN: Redleaf Press.

Rinaldi, C. (2006). *In dialogue with Reggio Emilia: Listening, researching and learning*. New York, NY: Routledge.

Rollnick, M., & Rutherford, M. (1993). The use of a conceptual change model and mixed language strategy for remediating misconceptions on air pressure. *International Journal of Science Education, 15*(4), 363–381.

Seidel, S. (2001). Understanding documentation starts at home. In C. Giudici, M. Krechevsky, & C. Rinaldi (Eds.), *Making learning visible: Children as individual and group learners* (pp. 304–311). Reggio Emilia: Reggio Children.

Sharapan, H. (2012). From STEM to STEAM: How early childhood educators can apply Fred Rogers' approach. *Young Children, 67*(1), 36–40.

Stake, R. (2000). Case studies. In N. Denzin & Y. Lincoln (Eds.), *Handbook of qualitative research* (2nd ed., pp. 435–454). Thousand Oaks, CA: Sage Publications.

Strauss, A., & Corbin, J. (1998). *Basics of qualitative research: Techniques and procedures for developing grounded theory* (2nd ed.). Thousand Oaks, CA: Sage Publications.

Thompson, C. (1995). "What should I draw today?" Sketchbooks in early childhood. *Art Education, 48*(5), 6–11.

Tippet, C., & Milford, T. (2017). Findings from a pre-kindergarten classroom: Making the case for STEM in early childhood education. *International Journal of Science and Math Education, 15*(Suppl. 1), S67–S86. doi:10.1007/s10763-017-9812-8

Tuttle, N., Kaderavek, J., Molitor, S., Czerniak, C., Johnson-Whitt, E., Bloomquist, D., Namatovu, W., & Wilson, G. (2016). Investigating the impact of NGSS-aligned professional development on preK-3 teachers' science content knowledge and pedagogy. *Journal of Science Teacher Education, 27*, 717–745. doi:10.1007/s10972-016-9484-1

Vasquez, J., Sneider, C., & Comer, M. (2013). *STEM lesson essentials, grades 3–8: Integrating Science, Technology, Engineering, and Mathematics*. Portsmouth, NH: Heinemann.

Whitin, P., & Whitin, D. (1997). *Inquiry at the window: Pursuing the wonders of learners*. London: Heinemann.

CHAPTER 7

Inquiry-Based Learning in Statistics: When Students Engage with Challenging Problems in STEM Disciplines

Theodosia Prodromou and Zsolt Lavicza

Abstract

This chapter reports on the analysis of the unstructured interviews of mathematics teachers who reflected on the classroom discussions between researchers, teachers and middle school students who engaged in critical and creative thinking during solving complex and authentic problems that require students to make meanings of the data from Science, Technology, Engineering and Mathematics (STEM) disciplines; promote discussions to deepen their statistical understanding; and enhance productive classroom norms for statistical inquiries. Outcomes of this research study include identification and illustration of classroom norms for statistical inquiries and facilitate students' inquiry-based statistical learning and teachers' planning for inquiry learning.

Keywords

STEM – inquiry-based – statistics – enquiry cycle

1 Introduction

Knowledge about data and its analysis is becoming increasingly important at all parts of life. Data is being continuously collected from everyone, analysed and fed back to people in various forms. This data is essential for governments and companies to make decisions about the future and assist people's lives. Thus, it is important to increase people's knowledge about data, its analysis and the related mathematical knowledge to support economic and social benefits in any countries.

Open data initiatives have been recently started around the world (United Nations, 2014) aiming to change the landscape of how data is captured, presented, and analysed as well as to further pinpoint the necessity of

 | DOI:10.1163/9789004391413_008

students and professionals to gain adequate statistical and mathematical literacies to be able to analyse and interpret data for future decision making. Hence, statistical and mathematical literacies are considered vital skills for not only prospective data users, but also for all citizens as they facilitate the interpretation and the critical evaluation of data-based arguments and statistical information processing (Gal, 2000), especially in Science, Technology, Engineering and Mathematics (STEM) disciplines. Currently, the term Big Data is being used to further accentuate the need to enable citizens to develop statistical skills, thinking and reasoning that are needed in both work and everyday life.

Mathematics along with statistics can be considered as a language for other STEM disciplines (Schmidt & Houang, 2007) because in the 21st century, statistical reasoning is commonly used to deal effectively with data and mathematical terms and visualisations are used to describe trends and patterns in huge amounts of information ("big data"). Moreover, workers across STEM disciplines use computational thinking and reasoning to communicate the solutions of problems while using powerful statistical software packages for mathematical modelling to make predictions about the relationships between different components of a system.

Although the terminology and skills acquired in Mathematics and Statistics are utilised in other disciplines, research and enrolment data show that students are losing interest in statistics and mathematics, often because their exposure to pedagogies and learning environments in these fields causes disengagement from learning (Mills & Goos, 2011). Among numerous attempts to reinvigorate pedagogies and learning environments, Inquiry-Based Learning (IBL) is one showing potential. IBL is a constructivist approach in which the overall goal is to engage and rebuild students' capacity to address complex problems and learn challenging mathematics through experimentation and discovery (ACOLA, 2013; Bruder & Prescott, 2013).

Australia's former Chief Scientist has commented that "when they do study [the sciences] at school … the best way to teach inspirationally is to teach it the way it's practiced" (Chubb, 2015). However, schoolteachers have rarely, if ever, experienced statistics in practice. Nor do they have experience in preparing students to integrate statistics into STEM disciplines, a major factor that is still considered a roadblock for many students from rural areas and culturally diverse backgrounds (Bonous-Hammarth, 2000; May & Chubin, 2003).

Statistics teachers' knowledge of the wide variety of uses of statistics in STEM is limited by the tradition of teaching statistics as a separate discipline: schools do not integrate statistics within the STEM syllabus or as integrated into the teaching of STEM subjects. The teaching of statistics is not taught in a way that highlights the relevance of the discipline to other areas of interest within STEM (Kember, Ho, & Hong, 2008; Howley, 2008).

The study described in this chapter is part of a bigger project that seeks to transform and improve statistical literacy in Australia, in line with the country's broader vision of supporting "informed and increased use of statistics" for representing, integrating and exploring complex information from diverse STEM sources in the Big Data Era (Prodromou, 2017).

This chapter reports on the experiences of middle school teachers implementing inquiry-based statistics pedagogy in middle school classrooms. This approach promotes curiosity, risk-taking and negotiation of statistical meanings. A particular focus is on expectations and norms for classroom behaviour in inquiry-based settings, especially those focused on having students work on group projects. This chapters analyses mathematics teachers' reflections on classroom discussions between researchers, teachers and students engaged in critical and creative thinking while solving complex and authentic problems that required the students to make meanings of the data from STEM disciplines, that promoted discussions to deepen their statistical understanding, and that enhanced productive classroom norms for statistical inquiries. This chapter identifies and illustrates classroom norms for statistical inquiries to facilitate students' inquiry-based statistical learning and teachers' planning for inquiry learning.

The project will theorise the role of inquiry-based education (IBE) and mathematical inquiry learning in deepening students' mathematical learning when they explore data from STEM disciplines.

2 Related Studies and Concepts

2.1 *Inquiry-Based Education*

Inquiry-based education (IBE) is becoming an increasingly important element of developing new curricula and pedagogical approaches. Artigue and Blomhoj (2013) summarised and conceptualised the theories and initiatives behind inquiry-based education. They reached back to the early 20th century ideas of the American educational philosopher John Dewey who stated: "Education should be for all, stimulate students' interest for learning and cultivate their autonomy, aim at the formation of human beings able to play an active role in the development of societies, and reject traditional teaching practices focusing on instruction and drill" (Artigue & Blomhoj, 2013, p. 298). This important review article examined theories and six frameworks[1] both for inquiry-based mathematics education (IBME) and inquiry-based science education (IBSE), which are both relevant for the proposed study.

Inquiry-based learning practices aim to engage students in constructing knowledge within communities of learners (Baron & Darling-Hammond, 2010; Franke et al., 2007; Goos, 2004; Hunter, 2012). Inquiry-based learning was overly structured or open inquiry learning ("discovery learning"), was considered to be ineffective (Confrey, 1991; Hatie, 2008). Makar and Fielding-Wells (2017) used the term "mathematical inquiry" to describe a process that uses mathematical evidence to address complex problems.

In mathematical inquiry, students deal with uncertainties and ambiguities and take intellectual risks to solve difficult problems or make difficult decisions whilst integrating information from complex problems and reworking or recomposing ideas with their classmates and with their teacher's support. Inquiry-based education requires students to discover knowledge and concepts, and to raise questions and ideas that sometimes are not even included in the content of their lesson plans.

Mathematical inquiry is related to positive learning benefits in mathematics education such as collaboration, improved communication, creativity and deep mathematical thinking (Barron & Darling-Hammond, 2010; Bruder & Prescott, 2013). These benefits enhance students' engagement and interaction in mathematical classroom practices. Students' interactions impact their mathematical thinking and "the way they are viewed as competent in mathematics, their ability to perform successfully in school" (Franke et al., 2007, p. 226). Students need to be encouraged to verbalise their thinking and justify their answers having educators taking simultaneously into consideration contextual factors of the STEM discipline.

The summarised framework for IBME includes ten recommendations (p. 13) for employing IBME in schools:

1. the 'authenticity' of questions and students' activity in terms of connection with students' real life and link with out-of-school questions and activities;
2. the epistemological relevance of the questions from a mathematical perspective, and the cumulative dimension of mathematics;
3. the progression of knowledge as expressed in the curriculum;
4. extra-mathematical questions and the modelling dimension of the inquiry process;
5. the experimental dimension of mathematics;
6. the development of problem-solving abilities and inquiry habits of mind;
7. the autonomy and responsibility given to students, from the formulation of questions to the production and validation of answers;
8. the guiding role of the teacher and teacher-student(s) dialogic interactions;

9. the collaborative dimension of the inquiry process;
10. the critical and democratic dimensions of IBME.

Moreover, the widely accepted framework from the USA National Science Education Standards (National Research Council, 1996, p. 4) recommends: (a) students create their own scientifically oriented questions; (b) students give priority to evidence in responding to questions; (c) students formulate explanations from evidence; (d) students connect explanations to scientific knowledge; and (e) students communicate and justify explanations. The PRIMAS project (http://www.primas-project.eu, 2011) formulated a broader version of inquiry-based teaching practice in science and mathematics (Maaß & Doorman, 2013) (see Figure 7.1)

Teachers also need to be confident when guiding students' discussions and collaborations maintaining students' intellectual risk-taking and deep mathematical thinking.

Makar and Fieldings-Wells (2017) argued that teachers need time and support to adopt mathematical inquiry pedagogies confidently and successfully. They pointed out that in some cases even with support, teachers may not be able to adopt mathematical inquiry approaches (Makar, 2011; Makar & Fielding-Wells, 2011) in their teaching practices because of the complexity inherent in inquiry practices. The success of adopting inquiry practices depends heavily on creating conducive classroom norms, which are shared understandings of the behaviors of both students and teachers and study the development of those norms through the interactions amongst students, tasks, and teachers (Goss, 2004) and the emergence of new praxeologies (Prodromou

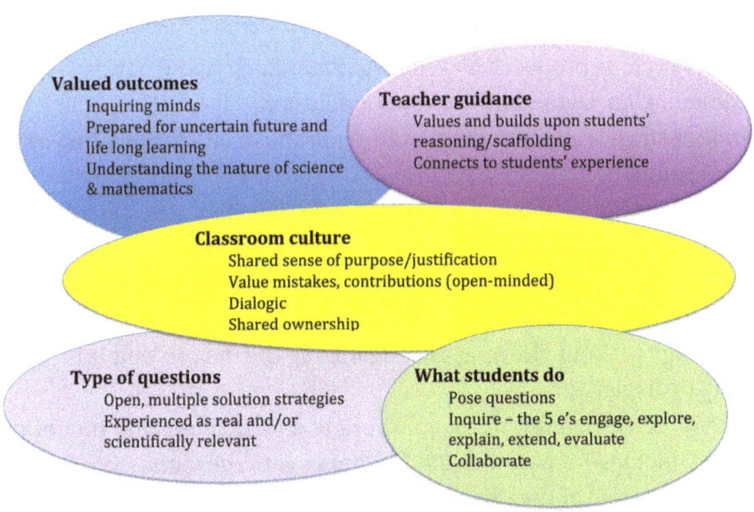

FIGURE 7.1 Essential ingredients in inquiry-based education (Artigue & Blomhoj, 2013, p. 801, reprinted with permission)

et al., 2017). Research conducted on classroom norms of mathematics has mainly focused on identifying norms of mathematics that are engaged in "instruction following an inquiry tradition" (Yackel, 1995, p. 131).

An interesting observation is that many of the examples that implement inquiry-based pedagogy typically focus on problems with relatively simple solutions. However, this research focuses on students' working with complex and challenging statistical problems in STEM disciplines. In the case of complex problems, students provide justifiable solutions drawing on various data sources available to them in the statistics enquiry cycle.

2.2 *The Statistics Inquiry Cycle*

"Statistical thinking is a general, fundamental, and independent mode of reasoning about data, variation, and chance" (Moore, 1998, p. 1254). Statistics is a branch of mathematics that deals with the collection, analysis, interpretation, presentation, and organization of data in an attempt to find the patterns and relationships of data. In contrast, the discipline of Mathematics deals with the exploration and use of patterns and relationships in quantities, space, and time.

Although these two disciplines use different ways of thinking and problem solving, they both equip students with means for investigating, interpreting, explaining, and making sense of the world in which they live. Conducting statistical investigations about STEM disciplines involves gathering information and seeking meaning from that information in order to learn more about phenomena, to inform decisions, to make statements about the data, and to answer questions. Statistical investigations are conducted using an inquiry cycle, which defines the way one acts and thinks about during the course of a statistical investigation (Wild & Pfannkuch, 1999).

The inquiry cycle as proposed by Wild and Pfannkuch (1999) consists of five stages: Problem, Plan, Data, Analysis, and Conclusion.

- The *problem* stage is about understanding a problem and formulating a statistical question.
- The *planning* stage is about deciding on how the data will be collected, measured and analysed. This stage also identifies sources of data: a primary source, i.e., data collected by students, or secondary sources, i.e., data already collected by someone else.
- The *data* stage is about cleaning and organizing the data so that it is useful and ready to display.
- The *analysis* stage is about using appropriate data displays and numerical summaries, looking for patterns and reasoning with the data.
- The *conclusion* stage of the cycle is about drawing conclusions and communicating the results of statistical investigations.

One of the advantages of the inquiry cycle according to Wild and Pfannkuch (1999) is that it provides more structure in the learning process. This study capitalizes on this advantage to investigate the development of classroom behavioural norms that facilitate students' learning of statistics when engaged in cycles of inquiry to explore complex and challenging problems in STEM disciplines.

3 Methodology

This study draws on research on inquiry-based learning and teachers' emerging pedagogies and experiences when adopting mathematical inquiry practices in teaching statistics and students' engagement with inquiry-based argumentation in statistics. The project aims to develop new knowledge about the classroom norms of statistical inquiry and does so by drawing on analysing the data of this research study.

In the first instance, the researcher of this project analysed the discussions amongst the researcher and the 3 secondary mathematics teachers who attended a professional development programme about inquiry. During this process they designed their lessons to promote the development of norms of mathematical inquiry in their classrooms. In this chapter, we will not look at the analyses of this data.

Chosen videos of the classroom inquiry lessons will be analysed to generate potential classroom norms of mathematical inquiry in statistics, including classroom discussions between the researcher, teachers and students during students' engagement with data from STEM disciplines. The overarching research question for this project is, "How do classroom norms of mathematical inquiry develop when middle school students are engaged with challenging statistical problems in STEM disciplines?" More explicitly, this research question would be addressed through the following sub-questions:

i. How do students reason and make meanings of the data from STEM disciplines when conducting statistical investigations in an inquiry-based classroom activity?
ii. How do teachers support the development of classroom norms of statistical inquiry during students' engagement with the statistics enquiry cycle and how they facilitate students' inquiry-based learning?

3.1 *Design, Participants and Tasks*

We will adopt design research methodology (Cobb et al., 2003) in this chapter, because "Design based research embraces the complexity of classroom settings and takes into consideration the unpredictability of classroom dynamics. In this project, we study the development of mathematical inquiry norms by: (a) iteratively studying the researcher's, teachers' and students' interactions during classroom inquiry lessons; (b) the researcher intervenes to advise teachers to initiate, maintain or develop norms of mathematical inquiry; (c) establish cycles of iteration, ongoing reflection and feedback to refine teaching practices and learning environments.

Participants were selected from several Australian secondary schools that are situated in rural areas. The team consisted of: (a) two researchers who had developed expertise in mathematical inquiry and students' reasoning about statistical concepts when engaging with the different stages of the statistical enquiry process; (b) three secondary mathematics teachers who adopted inquiry pedagogy as their teaching pedagogies after they attended the professional development programme about inquiry; and (c) 86 secondary Year 8–9 students who were taught by the 3 mathematics teachers.

4 Data Collection and Analysis

The data for this chapter were collected immediately after the professional development program about inquiry. The data include video recordings of the 3 secondary mathematics teachers' interviews immediately after their inquiry lessons and it was collected for 2 months. Teachers reflected on their classroom lessons immediately after the lessons, providing rich illustrations of examples of mathematical inquiry. They responded to questions about: (a) how students reason and make meanings of the data of STEM disciplines when conducting statistical investigations and follow the inquiry statistical cycle; and (b) their classroom inquiry norms. Each interview lasted for half an hour to 45 minutes.

All the interviews were transcribed and coded, focusing on observable norms; the strategies used in initiating, developing, and maintaining desired classroom inquiry norms; and how those norms supported students to make meanings of the data from STEM disciplines during their statistical investigations. The recordings of the interviews were then re-listened and the codes re-examined and discussed within the research team and organized into thematic groups.

4.1 *Findings*

Findings from three teachers in the critical experience reflections are presented below:

4.2 *Sue*

Sue is a veteran female teacher from Australia. After she attended the professional development programme about inquiry, she tried to change her teaching practice by implementing an inquiry-based learning approach in her classes, but she expressed that did not feel confident doing it.

In her classroom of 28 Year-9 students, she taught the Statistics Inquiry Cycle by engaging students in the creation of posters for the National Schools Posters Competition (NSPC) that required students to engage in project-based learning activities. In NSPC there are five divisions representing school stages 2 to 6, with small cash prizes ($50 to $200) awarded to winning teams and schools to motivate students' learning. Winners may be promoted to attend the biennial International Statistical Literacy Project competition. Project learning-based activities involved teams of 2 to 5 students creating an informative poster presentation (akin to a conference poster) addressing a practical question on an area of their interest in any field and following the Statistics inquiry Cycle. Sue asked one of the researchers to support her attempt and be present in the classroom to facilitate the activity's delivery and engage students with discussions while she taught every stage of the enquiry cycle. The teaching involved setting appropriate research questions as well as collecting, recording, presenting and interpreting data. When Sue taught each stage of the statistical cycle, she showed examples of posters to illustrate the stage. Students carried out each stage in practice before moving to the next stage.

For example, when Sue taught the first stage, she discussed with her students how to know whether they have formulated an appropriate or inappropriate statistical question. Afterwards, students worked in groups to understand different problems from STEM disciplines and articulated appropriate statistical questions in the classroom for classroom discussion. Researchers orchestrated those discussions and actively engaged students with authentic learning opportunities integrating statistics and STEM concepts. In particular, the researcher aimed to introduce concepts of different stages of the enquiry cycle, to students in an enjoyable, interactive, project-based manner.

The researcher in Sue's classroom drew upon STEM examples, for example, the impact of temperature on the bounciness of a ball; the effect of sugar on concentration and students' test performance; how brands change the perceived taste of foods; weekly use of plastic bags and sustainability; etc. These examples were chosen to encourage and inspire Year 9 students towards investigations of the surrounding environment and its sustainability and/or statistics in any field of interest, boosting students' confidence in being able to contribute to the development of scientifically-rigorous sustainability pursuits, helping them believe such efforts are within their reach and mental capacity.

After Sue had taught the planning stage of the statistical inquiry cycle about deciding on how the data will be collected and measured, the researcher encouraged the practical application of this stage with students. She supported students to either collect primary data from other classrooms in their school or from their friends via social media (Facebook or Twitter), or to use secondary data collected from the Australian Bureau of statistics (ABS) or Open Data (using the term "Open Data" for data "offered by several official statistics agencies, and any of their accompanying Online analytical processing (OLAP) facilities." Prodromou, 2017, p. 2). The researcher helped students to use Open Data in their poster projects illustrating the types of contexts that are available and their connections with the Australian school curriculum. Sue and the researcher worked with each group to assist in the recording of data and frequencies of categorical data in tables. The tables of each group were further shared and discussed with the entire classroom. Comparisons amongst tables of different groups were made and conclusions were verbalized. Students received similar support when they used appropriate visualization tools to create representations and develop the understanding necessary to analyse data and draw conclusions. The peer work in groups and then the following classroom sharing and critical discussions aimed to build appreciation of the practice of statistics, and further to assist students to become critical thinkers in judging the claims of their classmates (and, later, as statistically literate adults, claims made by others).

4.3 *Matt*

Matt is an early career teacher who participated in the project. After he had attended the professional development programme about inquiry, he spent time with his students to discuss how statistical, quantitative and scientific skills can help them to better understand the world. He showed them examples of interdisciplinary statistical investigations that needed practical statistical expertise to help students understand why STEM careers can be interesting and to pursue a university degree in STEM. He elaborated on why students need to understand statistical concepts and acquire statistical skills to access and succeed in higher education and to acquire the required skills for the Big Data era.

Matt drew upon the successes of experienced tertiary educators and practitioners, in order to encourage and inspire his students. He then showed statistical investigations to students that were conducted by statisticians and other students who had previously participated in the National Statistics Concepts Competition. He discussed the analytical skills and statistical competencies required for students to conduct statistical investigations and taught the statistical inquiry cycle. He encouraged students to set groups of 2–4 and discuss interesting topics for investigation in statistics, sustainability and the sciences. Matt monitored each group's discussion for every stage of the cycle. When all groups completed their work for each stage, they shared their work with all the other students as part of the classroom. Matt asked questions of the students that required demonstrating understanding of statistical concepts and critical thinking with respect to their project. Students had to verbalize and communicate their understanding of different aspects of their data. To develop educational techniques that inspire students to maintain their interest and engagement, it would help to use good classroom inquiry norms. Statistical investigations that were meaningful to students were those that used personal data of students or from their environments, for example: What subjects do students in years 3 and 4 prefer at school or what is Year 9 Weekly Computer Usage?

Matt successfully orchestrated the group discussions among team members, they shared and monitored discussions with the teacher, and they reflected their work with critical thinking.

4.4 *Judith*

Judith is a female in her 30s. After she attended the professional development programme about inquiry-based learning, she decided that it was necessary to initially set appropriate behavioral expectations and rules in her classrooms. In her classes, she discussed with all students how they were expected to behave towards each other and towards the materials they use in her mathematics classroom, setting expectations and rules. She explained in detail the importance of constructing mathematical understandings and knowledge when working collaboratively to experience a level of ownership, participate in instruction, and engage in mutually respectful and cooperative relationships. She then explained the importance of peer learning, the expectations of socially appropriate interactions among peers and interactions between the teacher and students.

Once students demonstrated a basic understanding of the core concepts of *trust, sharing, belonging* and *respect*, Judith and the students could jointly develop class norms that support the learning of concepts. Norms were then written at a specific level to not only specify the particular behaviours in which students are expected to engage, but also to make them applicable in a wide variety of situations in her mathematics classroom.

Once these classroom norms were developed, Judith introduced her Year 9 students to the National Schools Poster Competition (NSPC) and explained the aims of the project-learning activities. She set groups of students to work together and provided all groups with a list of possible topics of interest on which to conduct statistical investigations. Although the list of possible statistical investigations demonstrated the interdisciplinary nature of investigations, it also limited students' freedom in choosing engaging project-based investigations on their own. Groups were further given instructions that mainly focused on closed questions of problems that require all members of the group to collaborate and solve. Peer collaboration supports the sharing and strengthening of concepts students encounter during their engagement with the enquiry statistics cycle. Interactions of all groups with the teacher were supported and encouraged, but the interactions of different groups were minimal in classroom discussions.

5 Summary

The outlined practices of the three teachers showed us interesting ideas with regards to setting classroom norms. The presence of the researcher in Sue's classroom boosted her efficacy, facilitating the development of classroom norms as well as encouraging, inspiring and supporting Sue to understand the role of statistics in STEM practices. Also, it enabled her to incorporate statistics into her practices and connect to topics in higher education. Both Sue and Matt used only open questions that inspired discovery in the classroom, however, Judith utilised mostly closed questions, somewhat restricting the experimentation in her classroom. Nevertheless, Judith emphasised that students and the teacher jointly develop and implement classroom norms and shifted some of the responsibility for supporting and encouraging socially appropriate interactions from the teacher to students. It also helped to ensure that students better understood the expectations of the classroom community and provided a rationale for them to monitor and change their own behaviours.

6 Conclusion and Discussion

As explained in the beginning of the chapter, changing teachers' practices is difficult. Teachers new to inquiry-based mathematical and statistical teaching need substantial guidance in developing and implementing new teaching practices and classroom norms. Particularly, for shifting towards open, discovery-based, and experimental approaches, most teachers require confidence and thorough guidance. To develop practices through which they can inspire and encourage students to work thorough challenging ideas and investigate their own interests, use of the statistics enquiry cycle could offer a strong guidance to assist efforts to combine statistics and STEM ideas. This changing practice also involves the encouragement of group work; peer-collaboration and public sharing, which develops further understanding of relevant ideas and helps explore students' own interests. The experiences of the three teachers, and their self-reports showed that the presence and guidance of the researchers was crucial in their understanding the possible roles The experiences of the three teachers, and their self-reports showed that understanding the possible roles of statistics in STEM subjects and their connection to university studies and students' careers assisted teachers in their development of inquiry-based teaching. The project involved researchers who helped teachers to follow up their ideas, but it is not realistic to have researchers in all classrooms, thus, developing support systems would be important. During the training period the collaboration of teachers and researchers could have contributed to practice-oriented insights into classroom norms and pedagogies. In addition, the perspectives of practicing teachers, which differ from those of the researchers, could further adapt requirements of national curricula, by, for example, utilising the National Schools Poster Competition (NSPC) examples.

The NSPC is a valuable initiative that could motivate and engage students in STEM disciplines and demonstrate the interdisciplinary nature of investigations as well as offer ideas for utilising statistics and STEM together in an inquiry-based approach. The model, structure, resources and content for the NSPC provide teachers the means to implement inquiry-based learning while teaching the inquiry cycle for conducting statistical investigations. This will enable a larger number of students to become familiar with and interested in the analytical, statistical, and machine learning aspects of the Big Data era.

Note

1. (1) The problem-solving tradition, (2) the theory of didactical situations, (3) the realistic mathematics education programme, (4) the mathematical modelling perspective, (5) the anthropological theory of didactics in 1991, (6) the dialogical and critical approach to mathematics education.

References

Artigue, M., & Blomhøj, M. (2013). Conseptualizing inquiry-based education. *ZDM – The International Journal on Mathematics Education, 45*(6), 797–810.

Australian Academy of Science (AAS). (2015). *Desktop review of mathematics school education pedagogical approaches and learning resources*. Canberra: Australian Academy of Science.

Australian Academy of Science (AAS). (2016). *Mathematics by inquiry.* Retrieved from https://www.science.org.au/learning/schools/mathematics-inquiry

Australian Council of Learned Academies (ACOLA). (2013). *STEM: Country comparisons: International comparisons of Science, Technology, Engineering and Mathematics (STEM) education.* Melbourne: Australian Council of Learned Academies.

Australian Government. (2015). *Restoring the focus on STEM in schools.* Retrieved from http://www.studentsfirst.gov.au/restoringfocus-stem-schools-initiative

Barron, B., & Darling-Hammond, L. (2010). Prospects and challenges for inquiry-based approaches to learning. In H. Dumont, D. Istance, & F. Benavides (Eds.), *The nature of learning: Using research to inspire practice* (pp. 199–225). Paris: OECD.

Bonous-Hammarth, M. (2000). Pathways to success: Affirming opportunities for science, mathematics, and engineering majors. *Journal of Negro Education, 69*(1–2), 92–111.

Bruder, R., & Prescott, A. (2013). Research evidence on the benefits of IBL. *ZDM: The International Journal on Mathematics Education, 45*(6), 811–822.

Chubb, I. (2015). *Interview with 1233 ABC Newcastle.* Retrieved from http://www.chiefscientist.gov.au/2015/07/interview-1233-abc-newcastle/

Franke, M. L., Kazemi, E., & Batley, D. (2007). Mathematics teaching and classroom practice. In F. K. Lester (Ed.), *Second handbook of research on mathematics teaching and learning* (pp. 225–256). Charlotte, NC: International Age Publishing.

Gal, Y. (2000). *Adult numeracy development: Theory, research, practice.* Cresskill, NJ: Hampton Press.

Gonfrey, J. (1991). Learning to listen: A students' understanding of powers of ten. In E. von Glasersfeld (Ed.), *Constructivism in mathematics education* (pp. 111–138). Dordrecht: Springer.

Goos, M. (2004). Learning mathematics in a classroom community of inquiry. *Journal of Research in Mathematics Education, 35*, 258–291.

Hattie, J. (2008). *Visible learning: A synthesis of over 800 meta-analyses relating to achievement.* New York, NY: Routledge.

Howley, P. P. (2008). Keeping it real, keeping them interested and keeping it in their minds. *Journal of Statistics Education, 16*(1), 1–16. Retrieved from http://www.amstat.org/publications/jse/v16n1/howley.html

Kember, D., Ho, A., & Hong, C. (2008). The importance of establishing relevance in motivating student learning. *Active Learning in Higher Education, 9*(3), 249–263. doi:10.1177/1469787408095849

Maaß, K., & Doorman, M. (2013). A model for a widespread implementation of inquiry-based learning. *ZDM Mathematics Education, 45*(6), 887–899. doi:10.1007/s11858-013-0505-7

Makar, K. (2011). Learning over time: Pedagogical change in teaching mathematical inquiry. In J. Clark, B. Kissane, J. Mousley, T. Spencer, & S. Thornton (Eds.), *Proceedings of the Mathematics Education Research Group of Australasia annual conference*

(*Beth Southwell Practical Implications Award*). Alice Springs: Mathematics Education Research Group of Australasia.

Makar, K., & Fielding-Wells, J. (2011). Teaching teachers to teach statistical investigations. In C. Batanero, G. Burrill, & C. Reading (Eds.), *Teaching statistics in school mathematics: Challenges for teaching and teacher education* (pp. 347–358). New York, NY: Springer.

Makar, K., & Fielding-Wells, J. (2017). Shifting more than the goal posts: Developing classroom norms of inquiry-based learning in mathematics. *Mathematics Education Research Journal, 30*(1), 53–63.

May, G. S., & Chubin, D. E. (2003). A retrospective on undergraduate engineering success for underrepresented minority students. *Journal of Engineering Education, 92*(1), 27–39. doi:10.1002/j.2168-9830.2003.tb00735.x

Mills, M., & Goos, M. (2011). Productive pedagogies in the mathematics classroom: Case studies of quality and equity. In B. Atweh, M. Graven, W. Secada, & P. Valero (Eds.), *Mapping equity and quality in mathematics education* (pp. 479–491). Dordrecht: Springer.

Moore, D. (1998). Statistics among the liberal arts. *Journal of the American Statistical Association, 93*, 1253–1259.

National Research Council. (1996). *National science education standards*. Washington, DC: National Academy Press.

National Research Council. (2000). *Inquiry and the national science education standards*. Washington, DC: National Academy Press.

Pacific Policy Research Center. (2010). *21st century skills for students and teachers*. Honolulu: Kamehameha Schools, Research & Evaluation Division.

Prodromou, T. (2017). *Data visualization and statistical literacy for open and big data*. Hershey: IGI Global.

Prodromou, T., Robutti, O., & Panero, M. (2017). Making sense out of the emerging complexity inherent in professional development. *Mathematics Education Research Journal, 53*(1), 53–78.

Schmidt, W. H., & Houang, R. T. (2007). Lack of focus in the mathematics curriculum: A symptom or a cause? In T. Loveless (Ed.), *Lessons learned: What international assessments tell us about math achievement* (pp. 65–84). Washington, DC: Brookings Institution Press.

United Nations. (2014). *Open government and data services*. Retrieved from https://publicadministration.un.org/en/ogd

Wild, C. J., & Pfannkuch, M. (1999). Statistical thinking in empirical enquiry. *International Statistical Review, 67*, 223–265.

Yackel, E. (1995). Children's talk in inquiry mathematics classrooms. In P. Cobb & H. Bauersfeld (Eds.), *Emergence of mathematical meaning: Interaction in classroom cultures* (pp. 131–162). Hillsdale, MI: Lawrence Erlbaum Associates.

CHAPTER 8

Values in STEM Education: Investigating Macau Secondary Students' Valuing in Mathematics Learning

Chunlian Jiang, Wee Tiong Seah, Tasos Barkatsas, Sylvia Sao Leng Ieong and Io Keong Cheong

Abstract

Values are of paramount importance in our society and strengthening values in Science, Technology, Engineering and Mathematics (STEM) education has the potential to weave an interactive pattern between STEM and various societal structures so that these are continuously and critically examined. It could be argued that what students value in their mathematics learning steer their decisions and actions throughout the learning process. In this context, the 'What I Find Important' (WIFI) Study was designed to identify what students value in mathematics learning. Survey data collected from 612 Grade 8 students were analysed by means of Principal Components Analysis (PCA), showing that Macau students value *achievement; relevance; practice; technology; communication and development*. In this chapter we analyse and interpret these values in the cultural context of Macau. Differences between Macau students and their peers in the other three greater China regions (i.e., The Chinese Mainland, Hong Kong, and Taiwan) are also discussed.

Keywords

Conation – effective mathematics learning – factor structure – the Chinese learner – values

1 Introduction

Members of our society are experiencing cataclysmic changes that are taking place almost simultaneously. The role of the Anthopocene as a focal point in a multitude of philosophical, anthropological and sociological contexts, the significant contributions of digitalization, nanotechnology, synthetic Biology,

 | DOI:10.1163/9789004391413_009

space technology, medical advances, Cybernetics, M-theory, the Theory of Everything and the dominance of data analytic methods, the renewed interest in Science, Technology, Engineering and Mathematics (STEM) education – almost all mathematically oriented – are just a few and they have the potential to create many new perspectives and values. Indeed, Webb (2017) argued that:

> We live in a world that is in some sense mathematical – although in precisely what sense is still hotly debated by mathematicians, physicists, philosophers and others. We are as a result, innately mathematical beings. Crossing the road, catching a ball, stacking the dishwasher: in our everyday lives we are constantly, unconsciously, manipulating numbers, assessing shapes, calculating position, geometry and motion. (p. 1)

The authors of STEM and values: Creating values in science and technology education (Siemens Stiftung, n.d.), claimed that: "Values create an indispensable foundation for personal development and social cohesion by reconciling different convictions and serving as a guide. In addition, values strengthen identity" (p. 1). They also claimed that: "Linking STEM and values spreads specialized knowledge while providing pupils with a space for experimentation where they encounter, experience, and reflect upon values" (p. 1).

In a STEM context values are often neglected it could be argued that they are needed in many mathematical, scientific, engineering and technological contexts and applications to sensitise researchers, academics and other societal milieu on the impact research has on society, our planet and the survival of our species.

The whole push for STEM education reflects a concern that the society does not and – will not in the near future – have enough of these professionals, at a time when emerging technologies require more and more of these types of personnel and their expertise. Yet, many students are shunning away, selecting non-STEM areas of study both at secondary and university levels. What the issue might be that is preventing students to embrace STEM subjects – including chemistry, physics and advanced mathematics – might be some kind of internal barrier, which acts against the students adopting effective learning habits and dispositions. This is where values kick in – teaching students how to sufficiently value mathematics, science and technology subjects/units/courses, so that they are open to embracing all that cognitive and affective research studies have taught us. In this chapter's context, we would argue that STEM education research is mathematics education research and it is equally true that mathematics education research is STEM education research.

Members of our international research team have conducted a multitude of research investigations as part of the 'What I Find Important' (WIFI)

International Study, to investigate students' values and preferences in their mathematics learning. Studies of students' science and technology values and preferences will follow. In this chapter we will analyse and discuss the outcomes of the WIFI study that was conducted in Macau.

2 The Macau Context

Macau, a Special Administration Region of China (Macau SAR), has been a hub for cultural interflow and mingling between the West and the East for over 400 years. Therefore, the education system in Macau has creatively integrated the western and eastern traditions and taken on unique characteristics. Macau students' participation in the international 'What I Find Important (in my mathematics learning)' (WIFI) Study definitely provides insight into what really works for Chinese learners and what other interested countries can learn from a Chinese education system. The 19-nation WIFI Study was initiated to investigate what students find important in their mathematics learning (Zhang et al., 2015). In this paper we will first introduce mathematics education in Macau and the WIFI Study, followed by an analysis and discussion of the findings of a survey of 612 Grade 8 Macau students. A brief comparison to the results obtained from students of the other three greater China regions will also be presented.

3 Mathematics Education in Macau

With a long history of Sino-Western cultural interflows, Macau has developed into a region unique in many respects, including education in general and mathematics education in particular. Education in Macau is characterized by a diverse pluralistic system allowing the co-existence of various forms of education (Wang, 2009). For example, regarding the duration of schooling, the systems currently running are the traditional Chinese school system of 6+3+3 (6 years for primary education, 3 years for junior secondary and 3 years for senior secondary), the British school system of 6+5+2, the Portuguese school system of 4+2+6, and the Luso-Chinese/Sino-Portuguese school system of 6+5. As far as the medium of instruction is concerned, there are Chinese-medium schools, English-medium schools and Portuguese-medium schools. Besides, there are separate schools for boys and girls as well as co-educational schools. Clearly, the various schools provide more alternatives for parents and students. In terms of financial resources, there are government-financed schools, official schools, and private schools. Unlike the other greater China regions where public education is prevailing, private schools are the mainstay of Macau education (Wang, 2009). In total, there are 77 schools in Macau and 87% of them are actually private schools. In the academic year 2014/2015, for example, 71,521students were studying in schools, more than 96% of them were enrolled in private schools (DSEJ, 2014). Among private schools, there are different forms of financial support: church-run schools, non-religious schools, charity schools and traditional prestigious schools.

The Education and Youth Affairs Bureau of the Macau SAR Government (DSEJ, abbreviation in Portuguese), gives schools great freedom in determining and selecting teaching language, teaching materials and contents, and even their own curriculum and course structure. For example, to help Macau students to become international citizens, two international schools have been created in Macau, one offering courses following the Canadian curriculum, and the other following the British National Curriculum. Several schools in Macau have two sections, one offering courses in English and the other using Chinese as the language of instruction. The schools are given enough autonomy to develop school-based curricula to cater for students' individual and diversified needs.

However, the fact that the private schools have the freedom to develop their school-based curriculums does not mean that the DSEJ gives up its governing role. In fact, the Education Reform Committee and the DSEJ had several laws promulgated to regulate curriculum designs for basic education from preschool to high school levels even before Macau was officially reverted to China in 1999. For example, in 1991, Education Law (Law No. 11/91/M) was promulgated as the legal framework for Macau SAR's education system. It has been regarded as a milestone in Macau's education history from non-commitment to active involvement in education matters (Leung, 2011). From 1991 to 1999, the government promulgated fifteen laws for funding, regulation and provision/ delivery of school education (Leung, 2011). Since the handover to China, the DSEJ has initiated a series of educational reforms to promote the development of Macau's basic education. For instance, in 2006, the Non-tertiary Education Law was promulgated, aiming at protecting the rights of children to fifteen years (K-12) to free obligatory education. Within ten years, the Macau government's investment in non-tertiary education increased from MOP10.07 (USD1.27) billion in 2002 to MOP32.92 (USD4.12) billion in 2012 (Xinhua News, 2014). With the financial support from the DSEJ, the class size has been reduced from more than 45 to 25–35 (Vong, 2013). An education development fund has been set up to promote various educational programs and activities to achieve quality education for all in the 21st century. The programs cover the development of Basic Academic Attainments (BAA) for major school subjects including mathematics, teacher professional development, co-curricular activities, and after-school activities for students.

Although private schools in Macau maintain their autonomy in school management and their own school curricula, the government is aware of the importance of a unified curriculum and public examinations for the improvement of its education system as well as its manpower capacity. Regarding mathematics, the DSEJ issued the first curriculum standard entitled "Macau Mathematics Syllabus (Trial)" in 1988, which was revised in 1999. Six years later, the DSEJ initiated the development of "Basic Academic Attainments in Mathematics for Primary Schools (BAAMPS)" (小學數學基本學力要求), which stipulated the basic proficiency in knowledge, skills, abilities, attitudes and values upon the completion of primary education. It was released in 2011 (DSEJ, 2011a) and was piloted at lower primary levels (Grades 1–3) in eight schools in the 2012–2013 academic year and at upper primary levels (Grades 4–6) in the 2013–2014 academic year (DSEJ, 2013). BAAMPS will be implemented in the school year 2016–2017 (Macau Special Administration Region (SAR), 2015) in Grades 1–3, and in all primary schools in the school year 2017–2018. The BAA in Mathematics for Secondary Schools was released in 2014 and piloted in 2014–2016 and will be implemented in school year 2016–2017. The implementation of the BAA in mathematics (BAAM) will have significant influence on mathematics curriculum and teaching practice. Especially noteworthy is the fact that the BAAM spells out the minimum standards at various stages instead of the "ceilings." The development of the BAAM is not intended to unify all the courses and the teaching materials in Macau. Instead, schools still have the autonomy to develop their school-based curricula based on their educational visions, missions, and students' abilities (Wong et al., 2015). Due to the lack of locally-published textbooks for local schools, most Chinese-medium schools actually adopt textbooks used in the Chinese Mainland, while most English-medium schools use textbooks from Hong Kong (Tang, 1999). The implementation of different mathematics curricula brings about different teaching practices in schools (Oliveira et al., 2015). However, many Macau teachers tend to teach mathematics in a traditional way emphasizing exercises with variations and controlling classroom activities although they also encourage students to be engaged in the process of learning (Huang & Leung, 2004).

In addition to setting up the minimum standards for attainments in different subjects at different grade levels, the DSEJ has also initiated the following three schemes for in-service teachers' professional development to effectively implement the curriculum guidelines: (1) Award Scheme on Instructional Design (ASID); (2) Pilot Scheme of the Elementary Curricula; and (3) Study Plan of Leading Teachers. ASID is a scheme to encourage in-service teachers to develop lesson plans as a teaching unit, or a course/program spanning a semester/whole academic year, to do action research in their own classrooms, and run open classes. The designs that are rewarded will be uploaded to the DSEJ website for other teachers to use (DSEJ, 2011b). It is hoped that teachers' teaching abilities will be enhanced during the process. Several schools were involved in the pilot scheme of BAAMPS. As a result, experts from Macau and the Chinese Mainland have been invited to provide academic support and teacher training programs on how to implement the BAAM on the basis of whole semester/year schemes of work. They also observe teachers' classroom teaching with follow-up suggestions for further improvement. Workshops and sharing sessions are organized from time to time for in-service teachers. Forty leading teachers including heads of mathematics departments, school leaders, and experienced mathematics teachers were selected for a professional development program consisting of topic studies, curriculum studies, workshops, school visits, and sharing sessions, with the objective of solving problems such as the lack of coherence and continuity as found in the European in-service teacher education (OECD, 2005, 2009).

The DSEJ has not only funded programs for teachers' professional development but has also funded two programs to provide immediate help for students' continual development in all the subjects including mathematics. The DSEJ provides funding for schools to run after-school classes, allowing students to stay in school and complete their homework with assistance from teachers, which is called *Duke* (督课). In addition, the "Making a phone call" program is also ready to offer immediate help in the evenings. Trained student helpers are on call, giving some hints or guidance to help students solve their homework problems.

In addition, DSEJ has tried to enhance the development of education informatisation in order to achieve the equity of education through the following three steps (Educational Research and Recourse Department, DSEJ, 2017): (1) To equip IT resources to schools so that it reached the levels of developed countries/regions. For example, at the end of this stage, almost every classroom has been equipped with a set of computer, projector, and microphone system so that teachers could conduct multimedia-assisted teaching. (2) To set up the Education Development Fund from 2007 to support schools to update IT infrastructure including the teacher notebook computer program and IT education coordinator scheme (Fan, 2010), to organize training and communication activities so that all the stake-holders have the appropriate IT skills. During this stage, some schools set up e-learning platform through which teachers can do instructional and assessment activities as well as individualized tutoring work. (3) To explore how education and internet+ (e.g., cloud computing, mobile web, Internet of Things (IoT), big data, Artificial Intelligence (AI), etc.) can improve educational innovation and reform. Many new initiatives will be carried out to help students develop their characteristics and potential fully.

In summary, in Macau private schools – which enjoy high autonomy in many aspects including curriculum planning – far outnumber public schools. Macau teachers are encouraged to develop good lesson plans and are rewarded if their plans are evaluated with high scores by professors in higher institutions in the related subject areas. The DSEJ has released the BAAM to set the minimum standards for different stages and has also provided substantial support for its implementation at school level through in-service teachers' professional development programs. In addition to curriculum development and teachers' professional development, the DSEJ and the educators in Macau schools have also made great efforts to provide individualized instruction and help for students' learning. All these programs and efforts have brought about obvious achievements in education in Macau in the past 15 years, particularly in mathematics education. For instance, the mathematics achievement of Macau students in Programme for International Student Assessment (PISA) 2012 showed big progress in mathematics literacy, improving its ranking from 15th in 2009 to 6th in 2012 (Cheung et al., 2013).

However, even though Macau students have been doing relatively well in international mathematics achievement tests, this success has been achieved mainly by ongoing reviews and fine-tuning being made to the mathematics curriculum and by improving the quality of mathematics teachers. As the sociopolitical situation around us changes daily, the challenge for Macau educators and researchers to stay in the forefront of (mathematics) education is no less than that confronting their peers overseas. The construct of values in the context of mathematics education (Bishop, 1988) represents one of these new approaches to sustaining the high standard of school mathematics education in Macau.

4 Values in Mathematics Education

In the context of mathematics education, values might be considered to be "the deep affective qualities which education aims to foster through the school subject of mathematics" (Bishop, 1996, p. 19). Seah (2018) considered it differently:

> Valuing refers to an individual's embrace of convictions, which are considered to be of importance and worth. It provides the individual with the will and grit to maintain any 'I want to' mindset in the learning and teaching of mathematics. In the process, this conative variable shapes

> the manner in which the individual's reasoning, emotions and actions relating to mathematics pedagogy develop and establish. (p. 575)

Thus, we regard the "valuing" by students and teachers as a process of attaching different levels of importance to different facets of mathematics pedagogy as they experience it. Aspects of the mathematics discipline and how it can be effectively learnt/taught, in addition to more general aspects of learning and being, are important relative to one another in the minds of individual students and teachers. As outlined in the affective domain of the Taxonomy of Educational Objectives (Krathwohl et al., 1964), such valuing gives the individual a unique 'value-print' identity which characterizes the person, and which in turn affects how phenomena in mathematics pedagogy are interpreted, and how responses are structured. Such valuing can also be considered internalised within individuals and relatively stable compared with other affective variables.

There are two important implications arising from these characteristics of valuing. Firstly, we can have a deeper and more complete understanding of mathematics learners only by paying attention to what they value, since valuing affects cognitive functions and affective dispositions. Secondly, the mathematics classroom is a collective of individuals each with their own values. How do these personal values interact with one another in ways, which support and facilitate meaningful or productive learning of mathematics? What happens when conflicting values inevitably come together, and how do the outcomes affect students' developing values and in turn, their respective mathematics learning? In this context, no prior research appears to have been conducted to evaluate what Macau students value in mathematics learning. Macau's participation in the WIFI Study is thus an attempt to facilitate an evidence-based investigation into the convictions held by Macau students, convictions, which are vital to effective mathematics learning.

The multinational nature of the WIFI Study also allows for meaningful comparisons between the Macau data and those of other participating economies. In particular, we are interested in how similar or different Macau students' valuing is compared with their peers in the Chinese Mainland, Hong Kong, and Taiwan. This knowledge should inform future rounds of curriculum reform in Macau. Furthermore, as students from these four regions take the top positions in the international mathematics achievement tests (Mullis et al., 2012; OECD, 2013), a study on what they value in their mathematics learning process may provide some insights for students and educators in western countries.

Indeed, 1,386 Grades 5–6 students in the Chinese Mainland, Hong Kong and Taiwan had taken part in the WIFI Study (Zhang et al., 2015). Analysis of their data revealed the surprising finding that across the three regions, the most important six values were identical, that is, Grades 5 and 6 students in these three regions valued (1) *achievement*; (2) *relevance*; (3) *practice*; (4) *communication*; (5) *ICT*; and (6) *feedback*. The cross-cultural comparison also revealed, unsurprisingly, differences in the three regions. First, whilst *achievement* and *ICT* were ranked first and sixth, the relative valuing of the other 4 attributes (i.e. relevance, practice, communication, and feedback) varied. Second, an examination of the Principal Component Analysis (PCA) components within each economy indicated that while the same attributes might be valued by students in different economies there were subtle differences across cultures. For example, students in two or more cultures may value *achievement*, but if this involves memorizing facts (item 14) in one culture, it does not necessarily mean that memorizing facts is also part of the valuing of *achievement* elsewhere.

Yet, we cannot extrapolate the findings from these three Greater China regions to the Macau context. Therefore, Macau's participation in the WIFI Study allows us to understand Macau students' valuing, and to find out how its unique colonial heritage and the casino culture may have influenced its students' mathematics learning. In this context, it is also hoped that the results of this study will facilitate an analysis of how Macau students value their mathematics learning compared with their peers in the other three Greater China regions, and what the implications are for promoting effective mathematics learning in Macau schools.

5 Research Question

The research question guiding this Macau study is the following:

> What do Macau students value with regard to their mathematics learning?

6 Research Methods

6.1 *Sample*

The participants were 612 Macau Grade 8 (M = 189, F = 423) students from six private schools. Government and private schools had been invited to participate in the study but eventually only six private schools agreed to participate. Four of the schools are co-educational while the other two are single-sex girls' schools. Also, five of the schools use Chinese as the language of instruction and the sixth school use English as the language of instruction. Four of the participating schools had adopted mathematics textbooks from the Chinese Mainland, and the other two were using Hong Kong mathematics textbooks. Although the resulting sample is not representative the participating schools varied in terms of gender composition, language of instruction and mathematics textbooks used.

6.2 *Instrument*

A questionnaire designed and validated for the WIFI Study (Zhang at al., 2015) has been used for this study. The questionnaire consists of four sections: (A) Sixty-four 5-point Likert scale items; (B) 10 continuum dimension items; (C) 4 items requiring open-ended, scenario-stimulated responses; and (D) students' demographic and personal information. This paper reports the findings from Section A, where students were asked to evaluate the importance of each statement presented, using the 5-point Likert scale with 1 indicating "absolutely important" and 5 "absolutely unimportant." In other words, a lower score will correspond to a greater valuing.

6.3 *Data Collection*

The WIFI questionnaire was administered by the students' mathematics teachers in their respective mathematics classes. The questionnaires were collected immediately after the students had completed them.

7 Results

An initial data screening was carried out to test for univariate normality, multivariate outliers, homogeneity of variance-covariance matrices, multicollinearity and singularity. Descriptive statistics normality tests showed that assumptions of univariate normality were not violated. The questionnaire items were subjected to a Principal Component Analysis (PCA) by using SPSSwin. Reliability tests were also conducted.

7.1 *Principal Component Analysis (PCA)*

For the Principal Component Analysis, the significance level was set at .05, while a cut-off criterion for component loadings of at least .45 was used in interpreting the solution. The rotation method was a Varimax with Kaiser Normalization. The rotation converged in 6 iterations. The resulting rotated matrix consists of six components as shown in Table 8.1, each with eigenvalue greater than one, which explains 49.7% of the total variance, with 15% attributed to the first component – *achievement.*

The naming of the *six components* was guided by the nature of the questionnaire items associated with each component, which will be discussed below. This resulted in the following six value components: *C1: Achievement, C2: Relevance, C3: Practice, C4: Technology, C5: Communication* and *C6: Feedback.*

TABLE 8.1 Rotated component matrix

	Component					
	1	2	3	4	5	6
Component 1: Achievement						
Q50 GettingTheRightAnswer	.731					
Q49 ExamplesToHelpMeUnderstand	.716					
Q58 KnowingWhichFormulaToUse	.684					
Q56 KnowingTheStepsOfTheSolution	.680					
Q55 ShortcutsToSolvingAProblem	.654					
Q51 LearningThroughMistakes	.627					
Q63 UnderstandingWhyMySolutionIsIncorrectOrCorrect	.606					
Q54 UnderstandingConceptsProcesses	.601					
Q28 KnowingTheTimesTables	.590					
Q48 UsingConcreteMaterialsToUnderstandMathematics	.564					
Q42 WorkingOutTheMathsByMyself	.546					
Q52 HandsonActivities	.525					
Q13 PractisingHowToUseMathsFormulae	.511					
Q5 ExplainingByTheTeacher	.468					
Q47 UsingDiagramsToUnderstandMaths	.444					
Component 2: Relevance						
Q18 StoriesAboutRecentDevelopmentsInMathematics		.777				
Q17 StoriesAboutMathematics		.769				
Q61 StoriesAboutMathematicians		.708				
Q40 ExplainingWhereTheRulesFormulaeCameFrom		.668				
Q60 MysteryOfMaths		.614				
Q39 LookingOutForMathsInRealLife		.558				
Q11 AppreciatingTheBeautyOfMathematics		.513				
Q29 MakingUpMyOwnMathsQuestions		.451				
Component 3: Practice						
Q57 MathematicsHomework			.740			
Q62 CompletingMathematicsWork			.724			
Q36 PractisingWithLotsOfQuestions			.719			
Q43 MathematicsTestsExaminations			.661			
Q64 RememberingTheWorkWeHaveDone			.449			
Component 4: Technology						
Q23 LearningMathsWithTheComputer				.769		
Q24 LearningMathsWithTheInternet				.768		

(cont.)

TABLE 8.1 Rotated component matrix (*cont.*)

	Component					
	1	2	3	4	5	6
Q4 UsingTheCalculatorToCalculate				.699		
Q27 BeingLuckyAtGettingTheCorrectAnswer				.480		
Q25 MathematicsGames				.466		
Component 5: Communication						
Q7 WholeclassDiscussions					.704	
Q9 MathematicsDebates					.632	
Q3 SmallgroupDiscussions					.619	
Q1 Investigations					.526	
Component 6: Feedback						
Q44 FeedbackFromMyTeacher						.821
Q45 FeedbackFromMyFriends						.750
Q41 TeacherHelpingMeIndividually						.423

Further, the Kaiser-Mayer-Olkin (KMO) index is 0.92 and Bartlett's test of sphericity (BTS) is significant at the 0.001 level. Therefore, factorability of the correlation matrix was assumed. The Cronbach alpha reliability coefficients for each component were: *C1: 0.90, C2: 0.82, C3: 0.82, C4: 0.76, C5: 0.65 and C6: 0.64*, indicating high or acceptable levels of item reliability for all six components.

The six components derived from the Macau sample are almost identical to the components found in the other three Greater China regions (see Zhang et al., 2015). In each of the six components similarities as well as differences were found. We will use the six components derived in this study to discuss the similarities and differences between Macau students' valuing of mathematics and the China Mainland, Taiwan and Hong Kong students' (Zhang et al., 2015) valuing of mathematics.

Component 1: Achievement

The first component consists of 15 items, many of which emphasized the importance of getting correct answers (item 50) through examples (item 49), steps/processes/formulae (items 58, 54, 56), shortcuts (item 55), etc. Unlike the results obtained from primary school students in the other three Greater China regions, Macau secondary students do not think that memorizing facts (item 14) is important for their mathematics learning though they believe that knowing the times tables (item 28) is important. Macau students also think the use of diagrams (item 47), concrete materials (item 48), and hands-on activities (item 52) are important to their mathematics learning. We tend to think that students at secondary levels can deal with mathematics at an abstract level, however, Macau students still prefer to make use of concrete and/or less abstract representations to develop their mathematical understandings. "One problem multiple solutions" is one of the problem-solving activities that are valued by Chinese mathematics teachers (Cai & Nie, 2007), however, the Macau students did not value "Looking for different ways to find the answer" (item 15) and "Alternative solutions" (item 30) as the participants in the study of Zhang et al. (2015).

Component 2: Relevance

The second component consists of 8 items, which emphasized the development of mathematics, its mystery and beauty, and its applications in real life. The students think stories about recent developments in mathematics (item 18), stories about mathematics (item 17) and mathematicians (item 61) are all relevant to their mathematics learning. "Connection mathematics to real life" (item 12), "Explaining my solutions to the class" (item 19), "Mathematics Puzzles" (item 20), "Students posing mathematics problems" (item 21), and "Outdoor mathematics activities" (item 34) were included in the relevant component of the Zhang et al.'s (2015) study but not in the current Macau study.

Component 3: Practice

The third component consists of 5 items, all of which are related to the practice work in mathematics learning. Students not only think that completing work (items 57 and 62) is important but also think that a lot of practice is important (item 36). Mathematics tests/examinations are regularly arranged in Macau schools, which may lead Macau students to see them as part of their practice work in mathematics. In their preparation for the tests, some Macau students may try to remember the work that have been done (item 64), but this was not included in the study of Zhang et al. (2015).

Component 4: Technology

The fourth component consists of 6 items, most of which are related to learning technology including computer (item 23), internet (item 24) and calculators (items 4 and 22). Nowadays students often play games online using a computer or a smartphone. In Macau, (known as the Monte Carlo of the Orient), parents do not have much time to manage their children's activities at home as those in other Greater China regions, the students may actually spend more time playing games (item 25) than their counterparts in the other three Greater China regions. Therefore, it is reasonable for them to consider it a good luck to get the correct answer (item 27). The last two items are not included in the study of Zhang et al. (2015).

Component 5: Communication

The fifth component consists of 4 items, three of which are related to discussion (items 3 and 7) and debates (item 9). The fourth item is investigations (item 1), which often in Macau is conducted in groups where discussion and collaboration are involved. "Relating mathematics to other subjects in school" (item 10) was included in the relevant component of the Zhang et al. (2015) study but not in the current Macau study.

Component 6: Feedback

The last component consists of 3 items, two of them being related to the feedback from teachers (item 44) and friends (item 45), similar to the study of Zhang et al. (2015). However, for Macau students, one additional item focusing on the individualized help (item 41) is included in this component. It could be argued that Macau students regard individualized help as a form of gaining feedback about their learning progress.

7.2 *Multivariate Analysis of Variance* (MANOVA)

In order to investigate statistically significant differences by gender, age and school setting for each of the six derived components, a Multivariate Analysis of Variance (MANOVA) was conducted. The dependent variables (DVs) were the 6 components derived from the PCA and the independent variables (IVs) were *gender, school setting, and age*. The marginal means of the six components by gender, school setting and age are reported in Table 8.2.

An initial data screening was carried out to test for univariate normality, multivariate outliers (Mahalanobis' distance criterion), homogeneity of variance-covariance matrices (using Box's M tests) and multicollinearity and singularity (tested in the MANOVA analysis). Descriptive statistics normality

TABLE 8.2 Estimated marginal means of the six PCA components by gender, school setting, and age

	Gender		School setting		Age		
	Male	Female	Co-educational	Girls only	13-Year old	14-Year old	15-Year old
C1 Achievement	27.999[a]	29.073	28.362	29.420[a]	28.351[a]	28.300[a]	29.494[a]
C2 Relevance	19.909[a]	23.866	21.043	25.554[a]	21.793[a]	22.652[a]	23.195[a]
C3 Practice	12.309[a]	12.250	12.112	12.584[a]	12.074[a]	12.275[a]	12.460[a]
C4 Technology	14.023[a]	14.046	14.047	14.020[a]	14.168[a]	13.786[a]	14.160[a]
C5 Communication	8.874[a]	10.038	9.217	10.517[a]	9.304[a]	9.619[a]	10.028[a]
C6 Feedback	6.672[a]	7.160	6.727	7.540[a]	6.853[a]	7.172[a]	6.967[a]

a. Based on modified population mean

tests (normal probability plot, detrended normal plot, Kolmogorov-Smirnov statistic with a Lilliefors significance level, Shapiro-Wilk statistic, skewness and kurtosis) showed that assumptions of univariate normality were not violated. Mahalanobis' distance was calculated and a new variable was added to the data file. There were fewer than twenty outlying cases, which is acceptable in a sample of 612 students. These outliers were therefore retained in the data set. Box's M Test of homogeneity of the variance-covariance matrices (which tests the null hypothesis that the observed covariance matrices of the dependent variables are equal across groups) was not significant at the 0.001 alpha level and we therefore concluded that we have homogeneity of variance. The tests of between-subjects effects are reported in Table 8.3.

7.2.1 Gender

This section reports the results of the MANOVA analysis of the six PCA components by *gender*. All 612 students (M = 189, F = 423) declared their gender and completed all the items.

TABLE 8.3 Tests of between-subjects effects

IV	DV	Type III sum of squares	df	Mean square	F	Sig.	Partial eta squared
Gender	C1 Achievement	36.500	1	36.500	.507	.477	.001
	C2 Relevance	355.637	1	355.637	9.764	.002	.016
	C3 Practice	10.725	1	10.725	.549	.459	.001
	C4 Technology	.161	1	.161	.009	.925	.000
	C5 Communication	32.425	1	32.425	5.074	.025	.008
	C6 Feedback	.823	1	.823	.121	.728	.000
School Setting	C1 Achievement	38.648	1	38.648	.537	.464	.001
	C2 Relevance	912.497	1	912.497	25.053	.000	.040
	C3 Practice	35.798	1	35.798	1.831	.176	.003
	C4 Technology	.214	1	.214	.012	.914	.000
	C5 Communication	73.389	1	73.389	11.485	.001	.019
	C6 Feedback	46.035	1	46.035	6.780	.009	.011
Age	C1 Achievement	85.384	2	42.692	.593	.553	.002
	C2 Relevance	137.179	2	68.589	1.883	.153	.006
	C3 Practice	17.788	2	8.894	.455	.635	.002
	C4 Technology	20.674	2	10.337	.564	.569	.002
	C5 Communication	24.293	2	12.147	1.901	.150	.006
	C6 Feedback	10.429	2	5.215	.768	.464	.003

Mauchly's test of Sphericity was used to test if the error covariance matrix of the orthonormalised transformed dependent variable is proportional to an identity matrix. Because Mauchly's test for school setting is not significant, the assumption of sphericity is not violated. Levene's test of equality of error variances has been used to test for homogeneity of variance for each of the dependent variables. The tests indicate that homogeneity has not been violated for any of the six components.

Pillai's Trace criterion was used to test whether there are no significant group differences on a linear combination of the dependent variables. Pillai's Trace criterion is considered to have acceptable power and to be one of the most robust statistics against violation of assumptions. Since the multivariate effect for age is significant ($p < 0.05$), we have to interpret the univariate between-subjects effects by adjusting for family-wise or experiment-wise error using a Bonferroni-type adjustment, and we derive the adjusted alpha level 0.008 (0.05/6) (Coakes & Ong, 2011).

Using this alpha level, we have significant univariate main effects for the following variable:

– *Component 2: Relevance* [$F(1, 612) = 9.764$, $p < 0.001$, $\eta^2 = .016$]

Effect sizes were calculated using eta squared (η^2). In our interpretation of effect sizes, we have been guided by Cohen, Manion, and Morrison's (2007) proposal that 0.1 represents a small effect size, 0.3 represents a medium effect size, and 0.5 represents a large effect size. In our case, the effect size was .016 (small effect). An examination of the estimated marginal means for *Component 2* indicates that male students valued *mathematics relevance* more than female students.

7.2.2 School Setting

This section reports the results on the MANOVA conducted on the six components by school setting. All students (364 Co-Ed students and 248 Girls' schools students) completed all items.

Mauchly's test of Sphericity for school setting is not significant and there is no violation of the sphericity assumption. Levene's test of equality of error variances indicates that homogeneity has not been violated for any of the six components. Given that Pillai's Trace criterion for gender is significant ($p < 0.05$) we used a Bonferroni-type adjustment and the derived adjusted alpha level is: 0.008.

Using this alpha level, we have significant univariate main effects for the following variables:

– *Component 2: Relevance* [$F(1, 612) = 25.053$, $p < 0.001$, $\eta^2 = .040$]

The effect size, calculated using eta squared, was .040 (small effect). An examination of the estimated marginal means for *Component 2* indicates that students from Co-Educational schools valued *mathematics relevance* more than students from Girls' schools.

– *Component 5: Communication* [$F(1, 612) = 11.485$, $p < 0.001$, $\eta^2 = .019$]
The effect size, calculated using eta squared, was .019 (small effect). An examination of the estimated marginal means for *Component 5* indicates that students from Co-Educational schools valued *communication* more than those from Girls schools.

7.2.3 Age

No statistically significant differences were found between any of the six PCA components and the age variable.

8 Discussion and Conclusion

The survey of 612 Macau secondary mathematics students has suggested that these students valued *Achievement, Relevance, Practice, Technology, Communication,* and *Feedback.* Gender difference was only significant for the valuing of *relevance*, in favour of the boys. Students from co-educational schools also appeared to value *relevance* and *communication* more than their peers in girls' schools. No age difference was detected. These results have been found to be both similar to – and different from – those of the studies conducted in the Chinese Mainland, Hong Kong and Taiwan (Zhang et al., 2015).

The PCA components in those two studies share many similarities with students from the four greater China regions valuing *achievement, relevance, practice, technology, communication,* and *feedback.* Students from all four regions view *achievement* as the dominant component for their mathematics learning. However, unlike the students from the other three greater China regions, Macau secondary students tend to value *technology* more than the primary students from the other three regions. The participants of this study were secondary students who may have had more access to using computers at home than primary students, whereas primary school students were also the participants in the study of Zhang et al. (2015). The Macau government has invested generously in upgrading school IT facilities since 2001. Macau schools allow students to use the internet to search for resources they need, to submit their assignment, etc. especially when schools are closed due to bad weather like strong storms and typhoons. In Macau, it is not strange to see that a 5-year-old child has a smartphone or iPad and could use it relatively well. Indeed, IT has become part of the resources and toolkits that children can use.

In concert with the students from the other three regions, Macau students believe that good performance in mathematics can be achieved through pedagogical practices such as teachers' explanations, practice of mathematics formulae, use of examples, etc. Different from the students of the other three regions, Macau students prefer concrete materials, diagrams, and hands-on activities as facilitators of their mathematics learning, but they do not think that looking for alternative solutions is important. "One problem, multiple solutions" is one of the problem solving approaches that are highly valued by Chinese mathematics teachers (Cai & Nie, 2007). This instructional approach could lead to improvements in students' problem solving performance (Lee, 2011). However, Macau students may not have had enough experiences using this method in their mathematics learning (Jiang, Seah, & Cheong, 2017). Jiang at al. (2017) analysed the worked examples included in all the eleven teaching designs that won prizes in ASID 2008–2013 and found that multiple solutions were provided in only 6 out of the 533 problems (1.13%).

Browsing through the items that loaded on the second component (*relevance*) and noting that items 12 and 34 did not load on that component, it could be argued that what Macau mathematics students learn at secondary schools are considered detached from real life. However, this does not prevent them from performing well in PISA (Cheung et al., 2013), which tests students' abilities to apply what they learned in schools to solve real-world problems. It is well known that a plethora of mathematics educators consider real-world applications to be important for the development of students' mathematics proficiency (e.g. Bevil, 2003; OECD, 2013; Van den Heuvel-Panhuizen & Drijvers, 2014). An interesting question arises from this discussion. Which factors contribute to Macau students' above average overall performance in PISA 2012, even though they generally do not regard school mathematics as being applicable? Indeed, in PISA 2012 (OECD, 2013), the indices of students' exposure to word problems and applied mathematics for the four greater China regions were all below the OECD average, with Macau students being the second last on average, and Shanghai students the third last on average. However, the indices of students' exposure to formal mathematics for the four greater China regions were all above the OECD average, with Shanghai students' average being the best score and Macau students' average being the third best score. The learning of formal mathematics seems to have given Shanghai and Macau students an advantage over their peers elsewhere because the relationship between exposure to formal mathematics and performance is significant at the country, school, and student levels (OECD, 2013).

"Remembering the work we have done" is taken as important by Macau students, but not by those of the other greater China regions. In another small survey (Cheong, Jiang, & Seah, 2017), we asked students to include the importance level of remembering mathematical formulae, proof, definitions, theorems, etc. We found that, in additional to this list, Macau students thought that it is important to remember review problems as they prepared for quizzes/examinations, and problems where they often made errors but not problem types that were used by their teachers as examples, nor problem types they practised in classes and at home. Chinese learners often consider memorization as a positive attribute in the mathematics learning process although this may not be built upon conceptual understanding (Leung, 2001; Nie et al., 2010).

Living in a society that is surrounded by casinos and all sorts of gaming, Macau students do find mathematics games important for their mathematics learning and may believe that getting the correct answers is due to good luck. In a small survey on the games Macau students played and the effect of mathematics games on their mathematics learning, Jiang and Cheong (under review) found that Macau students often play Poker and magic cube, and they do believe that playing games helps them to develop their computational and other mathematical skills, and deep understanding. For example, about 64% of the grade 8 students thought that games made their mathematics learning easier, and about 70% thought that playing games helps them develop a deep understanding of mathematics contents (i.e., Poker in combination and permutations). It seems that playing mathematical games has become an important component of their mathematics learning experiences.

A unique valuing dimension reported by the students in Macau compared to their peers in mainland China, Hong Kong and Taiwan was that of *individualised help*. The importance of individualised help can be explained by a societal culture of after-school classes and "making a call for help." The latter is a programme offered by DSEJ to allow students to make telephone calls for help if they encounter problems with their homework at home. Trained student teachers will provide them with some help so that they can finish their homework. The learning of mathematics concepts is incremental, and thus, it is crucial to make sure that they understand what they are currently learning, especially those difficult content areas like algebra and Euclidean geometry. How do the after-school classes and "making a call for help" programmes help their mathematics learning is an aspect of the shadow curriculum that also needs to be further investigated?

No statistically significant gender differences were found in this study except for the valuing of *relevance*. Male students in our study indicated that relevance is more important for their mathematics learning than their female peers did. It seems that it is more important for boys that mathematics is relevant. Kaiser et al. (2012) explained that students' views about mathematics were influenced by gender-role stereotypes within society considering mathematics as a male domain. It is important to make girls aware that mathematics is relevant to their schooling as well as to their future life (Jacobsen, 2012). In fact, Macau schools may have set the same requirements for both boys and girls. For example, mathematics teachers in schools normally assign the same homework for boys and girls in the same class all the time. No wonder the results obtained from PISA 2012 (Cheung et al., 2013) show no difference in mathematical literacy in Macau's 15-year-old students although boys scored higher than girls across the nations participating in PISA 2012 (Stoet & Geary, 2013). It seems that Macau has achieved gender equity in mathematics performance, which is one of the objectives of mathematics education in many countries around the world (Else-Quest et al., 2010; Mullis et al., 2012).

Statistically significant differences between students from co-educational school settings and girls' schools with regards to the valuing of mathematical *relevance* and *communication* were found in this study. Students from girls' schools considered these aspects less important than those from co-educational schools. What learning experiences do girls obtain from their respective schools? To what extent are girls' school students' experiences different from their peers in co-education schools? Is it because they are not allowed to talk in mathematics classes too much or because they are too shy to communicate in mathematics classes? Further research is required to answer these questions.

While Macau students' mathematics achievement is amongst the best in the world, it is not as good as those of the Chinese Mainland, Hong Kong and Taiwan. One of the objectives of this study has been to contribute to this understanding. Our study here has revealed firstly that Macau students may not have had much experience with working with open-ended questions as well as with the applications of mathematics in real-world situations. Secondly, Macau students value not only basic mathematical facts (i.e., times tables) and formulae but also the work they have done, in particular, review problems for quizzes/ examinations, and problems where they often made errors. Thirdly, students in Macau appeared to attribute the attainment of correct answers to luck.

It could be argued that the diverse and pluralistic school system in Macau has provided its students with choices and alternatives, which is similar in some ways to – but also different from – their peers in the other three Greater China regions. This knowledge would also be useful for both Western academics and their peers elsewhere when they plan intervention programs in the future.

The integration of values in STEM education – mathematics being one of its four pillars – is still in its infancy but as the outcomes of this study suggest, it will be a worthwhile pursuit for educational systems around the world to pay attention to the roles played by this conative construct in facilitating STEM education, both in terms of how it might be best learnt and how the content might be best represented.

Acknowledgement

This project was sponsored by the University of Macau (UM) Multi-Year Research Grant. We express our gratitude for the support. Opinions expressed herein are those of the authors and do not necessarily represent the views of UM.

References

Bevil, A. P. (2003). *The effect of real-world mathematical applications on gifted and nongifted students' achievement and classroom learning environment* (Doctoral dissertation). University of Houston, Houston, TX.

Bishop, A. J. (1988). *Mathematical enculturation: A cultural perspective on mathematics education*. Dordrecht: Kluwer Academic Publishers.

Bishop, A. J. (1996, June 3–7). *How should mathematics teaching in modern societies relate to cultural values: Some preliminary questions*. Paper presented at the Seventh Southeast Asian Conference on Mathematics Education, Hanoi, Vietnam.

Cai, J., & Nie, B. (2007). Problem solving in Chinese mathematics education: Research and practice. *ZDM: International Journal on Mathematics Education, 39*(5–6), 459–475.

Cheong, I. K., Jiang, C., & Seah, W. T. (2017). What is more important to remember in mathematics learning. *Teacher Magazine, 57*, 61–63. [in Chinese]

Cheung, K. C., Sit, P. S., Mak, S. K., & Ieong, M. K. (2013). *Macao PISA 2012 report: Assessment of mathematical, scientific and reading literacy performance of 15-year-old students from an international comparison perspective*. Macau: Educational Testing and Assessment Research Centre, University of Macau.

Cohen, L., Manion, L., & Morrison, K. (2007). *Research methods in education* (6th ed.). London: Routledge.

Education and Youth Affairs Bureau (DSEJ). (2011a). *Educational statistics*. Retrieved September 27, 2016, from http://portal.dsej.gov.mo/webdsejspace/addon/allmain/msgfunc/Msg_funclink_page.jsp?msg_id-8525&

Education and Youth Affairs Bureau (DSEJ). (2011b). *Educational statistics*. Retrieved September 27, 2016, from http://portal.dsej.gov.mo/webdsejspace/addon/allmain/msgfunc/Msg_funclink_page.jsp?msg_id=8525&

Education and Youth Affairs Bureau (DSEJ). (2013). *小學教育課程先導計劃*. Retrieved September 27, 2016, from http://www.dsej.gov.mo/crdc/project/pilot_primary.html

Education and Youth Affairs Bureau (DSEJ). (2014). *Educational statistics*. Retrieved September 27, 2016, from http://portal.dsej.gov.mo/webdsejspace/addon/allmain/msgfunc/Msg_funclink_page.jsp?msg_id=8525&

Educational Research and Recourse Department (DSEJ). (2017). Education and internet+: Some insights. *Teacher Magazine, 57*, 2–4. [in Chinese]

Else-Quest, N. M., Hyde, J. S., & Linn, M. C. (2010). Cross-national patterns of gender differences in mathematics: A meta-analysis. *Psychological Bulletin, 136*(1), 103–127.

Fan, A. (2010). The readiness of schools of Macao to integrate IT in education and the extent of actual IT integration. *International Journal of Education and Development Using Information and Communication Technology (IJEDICT), 6*(4), 52–63.

Huang, R., & Leung, F. K. S. (2004). Cracking the paradox of the Chinese learners: Looking into the mathematics classrooms in Hong Kong and Shanghai. In L. Fan,

N. Y. Wong, J. Cai, & S. Li (Eds.), *How Chinese learn mathematics: Perspectives from insiders* (pp. 348–381). Singapore: World Scientific.

Jacobsen, L. (2012). Preface to "Equity in mathematics education: Unions and intersections of feminist and social justice literature." In H. Forgasz & F. Rivera (Eds.), *Towards equity in mathematics education: Gender, culture, and diversity* (pp. 31–37). Berlin: Springer Verlag.

Jiang, C., & Cheong, I. K. (under review). What roles do games play in Macao secondary students' mathematics learning? *Teacher Magazine.*

Jiang, C., Seah, W. T., & Cheong, I. K. (2017). Is one problem multiple solutions important? *Teacher Magazine, 56*, 69–74. [in Chinese]

Kaiser, G., Hoffstall, M., & Orschulik, A. B. (2012). Gender role stereotypes in the perception of mathematics: An empirical study with secondary students in Germany. In H. Forgasz & F. Rivera (Eds.), *Towards equity in mathematics education: Gender, culture, and diversity* (pp. 115–140). Berlin: Springer Verlag.

Krathwohl, D. R., Bloom, B. S., & Masia, B. B. (Eds.). (1964). *Taxonomy of educational objectives: Handbook II: The affective domain.* New York, NY: McKay.

Lau, S. P. (2009). *A history of education in Macao.* Macau: Faculty of Education, University of Macau.

Lee, S. Y. (2011). The effect of alternative solutions on problem solving performance. *International Journal for Mathematics Teaching and Learning.* Retrieved from http://www.cimt.plymouth.ac.uk/journal/lee.pdf

Leung, F. K. S. (2001). In search of an East Asian identity in mathematics education. *Educational Studies in Mathematics, 47*(1), 35–52.

Leung, J. Y. H. (2011). Education governance and reform: Bringing the state back in. In N. M. K. Lam & I. Scott (Eds.), *Gaming, governance, and public policy in Macao* (pp. 163–182). Hong Kong: Hong Kong University Press.

Macao Special Administration Region (SAR). (2015). 澳門特別行政區第*10/2015*號行政法規本地學制正規教育基本學力要求. Retrieved September 27, 2016, from http://bo.io.gov.mo/bo/i/2015/29/regadm10_cn.asp?printer=1

Mullis, I. V. S., Martin, M. A. O., Foy, P., & Arora, A. (2012). *TAMS 2011 international results in mathematics.* Chestnut Hill, MA: TIMSS & PIRLS International Study Center, Boston College.

Nie, B., Zheng, T., Sun, W., & Cai, J. (2010). 美国现代数学教育改革 [Current mathematics education reform in the United States of America]. Beijing: People's Education Press.

Oliveira, M. F., Negreiros, J. G., & Neves, A. C. (2015). A contribution from Macao, China, on teaching mathematics. *International Journal of Education and Research, 3*(11), 189–208.

Organization for Economic Co-operation and Development (OECD). (2005). *Teachers matter: Attracting, developing and retaining effective teachers.* Paris: Organization for Economic Co-operation and Development.

Organization for Economic Co-operation and Development (OECD). (2009). *Creating effective teaching and learning environments: First results from TALIS*. Paris: Organization for Economic Co-operation and Development.

Organization for Economic Co-operation and Development (OECD). (2013). *PISA 2012 results: What students know and can do – Student performance in mathematics, reading and science* (Vol. I). Retrieved from http://dx.doi.org/10.1787/9789264201118-en

Seah, W. T., & Andersson, A. (2015). Valuing diversity in mathematics pedagogy through the volitional nature and alignment of values. In A. Bishop, H. Tan, & T. Barkatsas (Eds.), *Diversity in mathematics education: Towards inclusive practices* (pp. 167–183). Switzerland: Springer.

Shen, H. (2002). Education and society in Hong Kong and Macau: Comparative perspectives on continuity and change. *Comparative Education, 38*(2), 239–245.

Siemens-Stiftung. (n.d.). *STEM and values: Creating values in science and technology education*. Retrieved from http://www.siemens-stitfung.org/en/projects/stem-and-values

Stoet, G., & Geary, D. C. (2013). Sex differences in mathematics and reading achievement are inversely related: Within- and across-national assessment of 10 years of PISA data. *PloS One, 8*(3), e57988.

Van den Heuvel-Panhuizen, M., & Drijvers, P. (2014). Realistic mathematics education. In S. Lerman (Ed.), *Encyclopedia of mathematics education* (pp. 521–525). New York, NY: Springer.

Vong, S. K. (2013, April 1). 澳門小班化教學的措施和現況 澳门日报2013年4月1日 [Class size reduction: Implementation and its current status in Macao]. *Macao Daily*. Retrieved June 3, 2016, from http://education.cloudmacau.com/news/news_detail.php?id=227#.V1GTwWdf2cM

Wang, Y. (2009). Preface 1. In S. P. Lau (Ed.), *A history of education in Macao*. Macau: Faculty of Education, University of Macau.

Webb, R. (2017). *Learn the language of reality. Infinity and beyond. Your ultimate guide to mathematics* [Editorial]. New Scientist the Collection, New Scientist Ltd, England.

Wong, I. N., Jiang, C., Cheung, K. C., Sun, X. (2015). Primary mathematics education in Macao: Fifteen years of experiences after the 1999 handover from Portugal to Mainland China. In X. Sun, B. Kaur, & J. Novotná (Eds.), *The twenty-third ICMI study: Primary mathematics study on whole numbers* (pp. 480–487). Springer.

Xinhua News. (2014). 澳门特区成立十五年来居民社会福利大步飞跃. Retrieved September 27, 2016, from http://news.xinhuanet.com/gangao/2014-12/09/c_127290262.htm

Zhang, Q., Barkatsas, T., Law, H. Y., Leu, Y. C., Seah, W. T., & Wong, N. Y. (2015). What primary students in the Chinese Mainland, Hong Kong and Taiwan value in mathematics learning: A comparative analysis. *International Journal of Science and Mathematics Education, 14*(5), 907–924. doi:10.1007/s10763-014-9615-0

CHAPTER 9

Perspectives on STEM Education in Preservice Primary Teacher Education

Wendy Nielsen, Helen Georgiou, Sarah Howard and Tricia Forrester

Abstract

Recent curriculum and regulatory changes in K-6 education require an integrative focus by primary teachers. Initial teacher education (ITE) responds to these changes with program innovation to support preservice teacher competencies, subject matter knowledge and pedagogical skill. STEM as a recent rhetorical focus provides new opportunity and impetus for ITE programs to support preservice primary teachers to integrate the STEM disciplines more deliberately. This chapter provides a number of examples of ITE program elements across the STEM Key Learning Areas that illustrate how preservice teachers can be positioned to take an integrative approach to science, technology and mathematics.

Keywords

Preservice primary teachers – science methods – STEM integration – initial teacher education

1 Introduction

Rhetoric around STEM has recently intensified, with stakeholders from business to education arguing for the need for current students (the future workforce) to develop sophisticated digital literacies and integrated STEM skills (Commonwealth of Australia, 2017; Department of Broadband, Communications and the Digital Economy, 2013). This rhetoric around STEM and STEM education positions Australian children as future innovators in the information age. However, STEM innovation and even digital literacy, to a degree, are nebulous terms and difficult outcomes to integrate into teaching and training programs, particularly in primary education, where a tension exists between depth and breadth.

An innovation agenda has driven science and technology education for many years, even as New South Wales has integrated these subjects at the primary level

 | DOI:10.1163/9789004391413_010

since the early 1990s (NSW Board of Studies [NSW BOS], 1991).[1] We argue that the rhetorical drive inherent in the 'STEM' agenda is not new, certainly not in NSW, which remains the only State in Australia that integrates science and technology in the K-6 syllabus (NSW BOS, 2012b). In theory, this positions NSW as a leader in STEM, at least from a curriculum perspective in the K-6 years. With the introduction of new national curriculum documents in Digital Technologies and Design and Technology (Australian Curriculum, Assessment and Reporting Authority [ACARA], 2015; NSW Education Standards Authority, 2017b, 2017c) and heightened focus on STEM as both pedagogy and discipline, there is new opportunity to consider cross-disciplinary or integrative perspectives in the STEM disciplines in initial teacher education. Teachers, of course, are constantly adapting to changes and initial teacher education (ITE) programs prepare primary preservice teachers to work across all subject disciplines in K-6 classrooms. However, with the complexity of the changes and pace of change this is increasingly difficult.

In this chapter, we outline approaches across the STEM disciplines in the primary ITE programs at the University of Wollongong in New South Wales, Australia. In designing subjects, we draw from theoretical perspectives in multimodalities (Jewitt, 2014; Kress, 2010), representational pedagogy (Tytler, Prain, Hubber, & Waldrip, 2014), mathematical reasoning (Ernest, 1994; Skemp, 1976) and scientific reasoning (Zembal-Saul, McNeill, & Hershberger, 2013) and in this chapter, use these to both ground and illustrate productive approaches to pedagogy in our primary ITE programs. We broadly conceive STEM as an integrative view of learning where the individual disciplines provide context for explorations across the fields. This conception is not based in critical perspectives on STEM, rather as a pragmatic response to the context of ITE in our setting and guidance from educational authorities in NSW (NSW Department of Education, 2017a). We focus our work in this chapter with the following question: *How does initial teacher education position primary teachers to implement an integrative focus in STEM?* We begin our discussion with a brief introduction to the current curriculum context in New South Wales particularly, with reference to the national curriculum in Australia more generally. These contexts form a backdrop for our descriptions of design decisions and strategies in our preservice primary teacher education programs where the subject areas are the location for PST explorations that cross the disciplinary boundaries of STEM.

2 Curriculum Context of STEM in Initial Teacher Education in New South Wales

Recent changes to the curriculum are pertinent to considering the contemporary context of STEM in primary teacher education in this chapter.

In Australia, delivery of educational services in the K-12 sector is a State and Territory responsibility, including development of syllabus documents and implementation support for teachers. However, in 2012, the national government introduced the Australian National Curriculum (ACARA, 2012) for Foundation[2] to Year 12 from which was intended to provide more consistency across the various jurisdictions. Each State or Territory could still exercise some control, with some adopting the Australian Curriculum wholesale and others developing or adapting it to create more detailed syllabus documents and resources for teachers.

The reality is that State authorities, primary teachers across Australia and teacher educators have been dealing with curriculum change for many years, both in terms of subject area content and change processes. For a variety of reasons, these significant curriculum changes have challenged relationships among national, state and local education authorities and service providers, including initial teacher educators and program providers.

Science as a curriculum area in the K-6 years in NSW includes the standard disciplines of physics, chemistry, biology and Earth sciences (NSW BOS, 2012b). The new national *Technologies* curriculum includes two strands: 'Design and Technologies' and 'Digital Technologies.' Design and Technologies corresponds roughly to what was 'Working Technologically' in New South Wales, which has been renamed in the new K-6 *Science and Technology* syllabus to 'Design and Production' (NSW Education Standards Authority [NESA], 2017b). With the range of technologies subjects in the later school years for NSW students, it could be argued that the NSW curriculum is more expansive than the National Curriculum. Even with this range, NSW has interpreted the national *Technologies* curriculum in the K-6 syllabus documents through the new strands of *Digital Technologies* and *Design and Production,* which in many ways complement the skills of the Working Mathematically (NSW BOS, 2012a). While the State level components do not map directly onto the two strands in the national curriculum, the *Science and Technology* syllabus has been revised to include aspects of the two new national *Technologies* strands (NESA, 2017b).

The NSW *Mathematics* syllabus has also undergone changes over time, but arguably, these have not been as dramatic as the changes to the *Science and Technology* syllabus. Mathematics curriculum documents in NSW have had a central focus on problem-solving approaches to teaching mathematics for nearly 30 years. Further, recommended teaching approaches have utilised challenging and rich mathematical tasks with real-life and cross-disciplinary connections across this time span (NSW BOS, 2002, 2012a; NSW Department of School Education, 1989). Further, these approaches align with research-based best practice and mirror the NSW Quality Teaching Framework (NSW Department of Education and Training, 2003).

The *Mathematics K-10 Syllabus* (NSW BOS, 2012a) not only details the content that needs to be covered in each stage of schooling, but also mandates the central importance of students working mathematically in learning this content: "Working Mathematically provides students with the opportunity to engage in genuine mathematical activity and develop the skills to become flexible and creative users of mathematics" (p. 36). There are five Working Mathematically components that underpin the delivery of content: Communicating, Problem Solving, Reasoning, Understanding and Fluency. Integrative learning experiences provide authentic contexts for students to experience mathematics as a problem-solving endeavour and creating opportunities for students to communicate substantively and to develop and share their mathematical reasoning, while improving conceptual as well as procedural understanding and fluency.

Changes to the official curriculum and syllabus documents impact program delivery in Initial Teacher Education (ITE) and provide impetus for examining course and program offerings. In addition to on-going curriculum change, a new accountability framework also includes Australian Professional Standards for Teachers (Australian Institute for Teaching and School Leadership [AITSL], 2011) where all teachers must satisfy accreditation requirements. Teachers graduating from an ITE program must demonstrate that they satisfy the Standards at the Graduate Teacher level and ITE program offerings provide opportunity for preservice teachers to develop the competencies progressively over the program. ITE programs are also subject to a national standards framework that is interpreted at the State level by the education authority (NESA, 2017a) and programs are accredited on a five-year cycle. Revisions to degree programs thus must satisfy each of the levels of education authorities, even when the review cycles rarely align. It is significant that official curriculum changes, Graduate Teacher accreditation requirements and ITE program level standards all impact on program delivery in ITE. Universities also have statements of Graduate Qualities that apply to any degree program. Satisfying each of these levels becomes very challenging when changes are continual.

In this ever-changing educational context, preservice teachers, through their ITE programs, must develop a range of competencies, subject matter knowledge and pedagogical content knowledge (Berry, Depaepe, & van Driel, 2012; Loughran, Mulhall, & Berry, 2008) to prepare for their work as classroom teachers. At the University of Wollongong, teacher educators who work in STEM fields take an integrative approach to subject design and delivery where preservice teachers gain knowledge and confidence as teachers of the STEM disciplines through direct experiences in the program.

3 Pedagogical Approaches to STEM in Initial Teacher Education

We view a focus on STEM as an opportunity for our preservice teachers (PST) to consider curriculum integration across disciplinary subject areas. This is particularly salient for future primary school teachers who are responsible for teaching across all school subjects. Considering how science can develop mathematical thinking or how technology education integrates reasoning and scientific process skills are important connections between the discipline areas of STEM. In this section, we review specific examples of strategies and pedagogies we use in the areas of STEM to highlight this integrative approach in our Initial Teacher Education programs for future primary school teachers.

3.1 *Science Education*

In primary science teacher education, PST gain experience with both science and technology when they learn about the official syllabus documents through lectures and tutorial activities. Their tutorial activities are structured, so PST learn about and practice the skills of 'Working Scientifically' and 'Working Technologically,' the key process skills in the NSW *Science and Technology* syllabus (NSW BOS, 2012b). Working Scientifically involves a range of skills and in conducting a first-hand investigation of their own choosing, PST work through what constitutes a 'fair test,' including how to manage variables, systematically record and analyse data and develop representations to display the data. In analysing the results of their investigations, PST utilize mathematical processes, develop reasoning skills and use common technology tools to present their analyses. With guidance, PST choose their own topics for investigation and recent choices include "The Five-Second Rule" (to see if food hitting the ground for five seconds grows more bacteria than food that is not dropped); if skim or whole cream milk used in making cupcakes affects how they rise; or how air pressure in a basketball affects its bounce. Developing the question for the investigation is scaffolded by teacher educators, as are methods for the investigation and techniques to collect data systematically. In gaining experience in 'working scientifically,' PST develop important background knowledge of science processes and gain confidence as future science teachers.

A second key task in *Science and Technology* in Initial Teacher Education involves 'Working Technologically' where PST design and produce a digital explanation of a science concept to explain the science to children. In this task, PST are assigned a syllabus outcome and supported in the design process to choose or make different representations of the science concept. The PST are advised to choose a topic area or concept within the assigned outcome that they judge to be difficult for children to understand. This supports the

PST to think deeply about the nature of the content knowledge in the syllabus outcome, but also to develop understanding of this content in order to explain it to others. In designing and developing a sequence of representations for the digital explanation, the PST use personal technologies and a range of media forms to integrate language, image, models, moving images and narration into a product that can be shared. Along with Nielsen, Hoban, and Hyland (2017), we call these products 'blended media' or 'educational mashups' (Hoban, Nielsen, & Hyland, 2016), which have evolved over time with the ease of use and availability of technology tools. This teaching strategy also provides a unique opportunity for PST to integrate their own understandings of science and technology content while using a variety of technology tools in a design process that creates a media product to share. Subject and activity design is generally informed by social constructivist views of learning and having PST develop multiple representations and work across multiple modalities is key to helping them develop science understandings.

3.2 *Mathematics Education*

Mathematics education subjects in initial teacher education are designed to engage preservice teachers as learners with a range of activities to develop relational understanding of the mathematics content, simultaneously providing them, as teachers, with models of best practice. Relational understandings underpin mathematical reasoning (Ernest, 1994; Skemp, 1976). Having experienced the powerful learning that comes from problem-solving approaches to teaching mathematics, these PST are then encouraged to reflect on and critically analyse the learning tasks and how they fit with the syllabus and research-based best practice.

Designing mathematics lessons, lesson sequences and programs in cross-disciplinary contexts is both modelled in our courses and specifically taught and assessed. First-year lesson planning tasks focus on embedding mathematics in cross-disciplinary contexts and through the semester, PST do several reflection tasks aimed at making these connections. We include two examples of such tasks here.

The first example illustrates a problem-solving, integrative experience that utilises readily available, quality resources in a tutorial activity. The activity for PST is derived from *The Case of the Mystery Bone – A Unit of Work on Measurement for Grades 5 to 8* (Clarke, 1996). This unit of work takes three to five weeks for children and includes a range of activities involving measurement of length and surface area; estimation; the use of percentages, decimals and fractions in real-life contexts; collecting and looking for patterns in data; notions of ratio and proportion; and fractions as operators. In the lessons for these activities, students become forensic scientists. In mathematics education, PST experience the activities firstly as children would, but then advance their understandings through pedagogical discussions.

In working with this unit on forensic science during tutorial activities, PST try to solve a murder investigation using the information presented in a series of newspaper articles reported over several days that gradually introduce new evidence. In small groups, PST conduct research and work with multiple forms of data and evidence to make predictions about the murder victim, sharing their strategies and reasoning with their peers. They decide which data they need to collect and how to use it to inform their predictions. They also develop reasoning as they gradually gain access to more evidence. The activities provide insights into the real-life applications of mathematics to forensic science in identifying victims and bringing perpetrators to justice. Following the activities, tutorial discussions ensue around the content covered, benefits for learning, class engagement, and links to learning theory and research-based best practice.

Preservice teachers have described this task as a benchmark in sourcing or creating their own lessons as well as developing deep understanding of the range of mathematical knowledges. They appreciate:

- the authentic science context where mathematics is used to solve real-world problems, highlighting the usefulness of mathematics outside of school;
- how the mathematics is embedded in a rich narrative that is highly engaging; and
- the clear coverage of essential syllabus content including: a range of mathematical concepts, measurement and calculation skills and all five processes of Working Mathematically.

In the second example, PST explore coding and robotics in preparation for going into a local school to assist with a coding and robotics workshop for students in Years 2, 4 and 6. PST use a range of online visual coding programs and apps in developing the skills to program robots to navigate a set course. These tasks foster computational thinking while providing meaningful, mathematically rich experiences that explore 2D and 3D space, position concepts and programming language and make strong linkages to the *Technologies* curriculum (ACARA, 2015).

Preservice teachers find these activities challenging but very engaging; one PST described them as "empowering." As their skills and knowledge increase, so too their confidence and readiness to further develop these experiences for use in their own future classrooms and even to contribute to the professional development of their colleagues in coming to terms with requirements to incorporate coding across the curriculum (NSW BOS, 2016).

3.3 *Technology Education*

In Initial Teacher Education, technology education focuses on the digital sections of the curriculum: digital media, digital technologies and information technology (NSW BOS, 2012b). These have typically been incorporated as part of the *Science and Technology* syllabus in NSW, and therefore taught as part of ITE *Science and Technology* methods subjects. However, with a revised focus on *Technologies* at the National level, the subject area is being reconsidered in the states and in ITE. A new syllabus for the subject area of technology is available in NSW for implementation in 2019. The aim of the digital component of the *Technologies* curriculum is to introduce students to technology systems such as networking, programming, databases, etc. to develop students' technology knowledge to operate in the modern world, but also to introduce them to the technical side of digital computing to support both their learning and their future work as teachers.

In ITE Technology subjects, PST engage with digital technologies through a 'learning to use and use to learn' format (Schank, Berman, & Macpherson, 1999) for two purposes: (i) to increase their technology knowledge and confidence with the knowledge; and (ii) to provide them with experience designing technology-based units of teaching. On the second point, they are specifically learning to use digital technologies and design technologically-integrated learning across subject areas. PST experience technologically-supported learning, such as coding or digital construction, and then use this knowledge to design similar learning experiences. This model accounts for PST having limited and/or very specific technology experience and broadens their understanding through use and experimentation, and two examples will be described.

The first is the development of a *Digital Scavenger Hunt*. The aim of the scavenger hunt is for PST to think about the affordances (Gibson, 1977) of digital technologies and how these enable and constrain mobility, data collection and experiential task design. A key aspect of using digital technologies is the selection and combination of different tools to perform a given task. This requires significant knowledge of the tool in order to assess its use. The scavenger hunt task has two parts. First, the lecturer designs an example scavenger hunt for PSTs to complete. To complete the hunt, students are given a list of 10 data items to collect, e.g. 'How many ducks are in the pond?' and 'How many people are in the Building 21 café right now?' and a tablet device. They are required to collect numeric data, e.g. frequency or an estimated measurement, and photographic documentation. Data are input to a Google Spreadsheet and/or Google Doc. Groups have limited time to collect the data, so they are advised to split up and collaborate through a Google Spreadsheet, a chat program and/or their mobile phones (and their own data plans). By way of follow-up and synthesis, they report on the data collection in a Google Doc and share with the rest of the tutorial group. A collective reflection on the experience discusses how the digital technologies were used, how they supported mobility, processes of data collection and task completion.

Students then create their own scavenger hunt, drawing on their previous experience. This requires students to work in small groups to design the learning experience using 8–10 different physical locations around the university, identifying data to be collected, and how and why it will be reported. The task design requires students to:

- work across multiple devices (e.g. a tablet, desktop/laptop computer, smartphone);
- consider the limitations of devices and networks (e.g. portability, data input, bandwidth, cloud storage, etc.) and how these affect learning design;
- construct a database fit for purpose (e.g. to allow team members to collect and share data from different locations); and,
- use different technologies to collect data, e.g. camera, spreadsheet, microphone.

A key element of creating the digital scavenger hunt is considering which tools best support mobile data collection and possibly other tools for working collaboratively in a group task environment. While PST enter ITE with a range of capabilities and experiences with technology tools, this problem-solving task involves negotiating the task and selecting and combining different technologies, which creates a much deeper understanding and confidence and reasoning to make and justify decisions.

The second example involves the creation of a digital portfolio. In the Technology subject PSTs need to keep records for a digital portfolio of their teaching unit development and completed tasks. To create the digital portfolio, they can choose among two cloud-based platforms: Google Drive or Microsoft OneNote. PST are given specifications of the portfolio and parameters for how it will be used in the subject through the subject learning management system, Moodle. The PST then need to choose the platform to use based on design decisions for their anticipated use of the portfolio. A demonstration portfolio is provided to the students to use as a model, one for each platform. Similar to the scavenger hunt, students consider the affordances of each platform and how it can fulfil the aims of the task. However, in this example, they must also consider their own work processes.

In both examples, students typically lacked deep knowledge or understanding of the digital technologies that they were being asked to use. By modelling use, such as a sample lesson and demonstration models, students can experiment and adapt to the task at hand. Importantly, these samples provide a scaffold for PSTs to develop confidence in both knowledge and performance of complex digital tasks. They can then reflect and draw on this experience when considering how to design for technology-based learning in their future work as teachers.

These descriptions of activities and design considerations for activities in ITE subjects across the STEM areas of science, mathematics and technology correspond to the official curriculum areas in both NSW and nationally. We have provided a brief overview of our efforts to provide integrative experiences and opportunities for primary PST. Further, because primary teachers must teach across all Key Learning Areas, their experiences during the ITE program are significant for developing their content knowledge, pedagogical skill and abilities to think and plan across the subject areas. This represents a key challenge both for them as developing practitioners and us as teacher educators responsible for the ITE program. As such, our programs could more clearly articulate connections across the STEM disciplines. As curriculum reform and ITE program redesign continues, increased effort to build integrative links is anticipated, which also opens new research opportunity for teacher educators. It may also be desirable to develop a new subject focused on STEM in order to make STEM more explicit at the program level, however, any new subject would necessarily replace something else, which thus opens other sorts of discussions.

4 Curriculum Change and STEM in ITE

While our respective subjects provide a context for PST to experience science, mathematics and technology as integrative, there are weaknesses to delivering them in different subjects. The risk is that PST continue to see the disciplines as separate and the knowledge, skills, tools and reasoning involved remain in disciplinary silos. Further, we need to do more to deliberately incorporate engineering. When our aim as teacher educators is that PST experiences in one place transfer across the program, we must continually work to better integrate STEM concepts and experiences across the curriculum in ITE. This is true for the STEM (and related) disciplines, but also for other subject areas. For example, while there is guidance in the State curriculum documents for how English or Creative Arts (NSW BOS, 2016) can include computational thinking, how teacher educators understand this guidance and integrate it into these subjects has yet to be explored. This is a challenge for all of us as teacher educators as we seek to deal with on-going curriculum change and accreditation requirements. ITE programs are accredited every five years on a cycle that does not map directly to the cycles of curriculum revision at the national or state levels. Thus, ITE programs are constantly trying to 'catch up' even as they are positioned to lead change through preparing the next generation of teachers.

As teacher educators, we constantly range through state and national education authorities' websites, seeking information and guidance about current and future changes. However, to anticipate change is risky as draft documents or commentary in early stages remain open to significant changes. While there is no national curriculum for teacher education, accreditation processes serve a similar function (among many others) to ensure that PST gain relevant content knowledge and pedagogical skill through experiences across the ITE program. And, subjects in ITE must reflect the KLAs in the school curriculum. At present, there is no school subject called 'STEM.' For teacher educators, interpreting official curriculum documents from national and state authorities means translating content for teaching children into content for teaching preservice teachers. A further complication in integrating STEM is that in this era of constant change, educators in ITE must be part of the conversation around changes and in response, develop meaningful learning experiences for PST so that they develop a depth of understanding about the nature and content of curriculum but also how to interpret the documents through successive cycles of change as they learn to respond confidently. STEM as a new rhetorical focus for curriculum is one such change.

We conclude the chapter with a call to our colleagues in primary preservice education to be deliberate in broadening views of STEM where disciplines provide contexts for methods and strategies in preservice primary teacher education. There are alternatives to teaching STEM across different subjects in ITE, even as such contexts provide opportunities for preservice teachers to seek out and build connections and reasoning across knowledge fields. This can effectively ground their preparation to teach across multiple knowledge fields and curriculum areas as K-6 teachers. Preservice teachers need such opportunities during initial teacher education in order to have a broad view of the nature of knowledge, but also to respond to the constantly changing environment in which they will work. They will be expected to be innovators in integrating curriculum areas as primary teachers but will also be required to adapt to curriculum change over time, which means responding to policy changes, accreditation authorities and impacts on children's learning.

Notes

1. The Board of Studies was the curriculum authority in New South Wales, but has undergone several restructures and name changes in recent years. The entity currently responsible for curriculum in NSW is known as the New South Wales Standards Authority (NESA). The government agency responsible for education

in NSW is the Department of Education, which has also undergone recent name changes and restructures, having previously been called the Department of Education and Training (DET) and the Department of Education and Communities (DEC).

2. 'Foundation' is the term used by the national authorities for the first year of formal schooling. In different jurisdictions across Australia, the year is also called "Kindergarten," "Early Stage 1" or "Prep." In NSW, the curriculum uses the term Kindergarten (or K) to refer to the first year of school, while Early Stage 1 is the name given to the phase of the syllabus used to plan for instruction for Kindergarten children.

References

Australian Curriculum, Assessment and Reporting Authority [ACARA]. (2012). *The shape of the Australian curriculum: Technologies*. Sydney: Australian Curriculum, Assessment and Reporting Authority.

Australian Curriculum, Assessment and Reporting Authority [ACARA]. (2015). *Digital technologies*. Retrieved from http://www.australiancurriculum.edu.au

Australian Institute for Teaching and School Leadership [AITSL]. (2011). *Australian professional standards for teachers*. Victoria: Education Services Australia.

Berry, A., Depaepe, F., & van Driel, J. (2016). Pedagogical content knowledge in teacher education. In J. Loughran & M. Hamilton (Eds.), *International handbook of teacher education* (pp. 347–386). Singapore: Springer.

Clarke, D. (1996). *The case of the mystery bone: A unit of work on measurement for grades 5–8*. Retrieved from https://www.mansw.nsw.edu.au/shop/books-middle-school/the-case-of-the-mystery-bone

Commonwealth of Australia. (2017). *Australia's national science statement*. Canberra: Government of Australia.

Department of Broadband, Communications and the Digital Economy. (2013). *Advancing Australia as a digital economy: An update to the national digital economy strategy*. Canberra: Australian Government.

Ernest, P. (Ed.). (1994). *Constructing mathematical knowledge*. London: Falmer.

Gibson, J. J. (1977). The theory of affordances. In R. Shaw & J. Bransford (Eds.), *Perceiving, action and knowing: Towards an ecological psychology* (pp. 67–82). Hillsdale, NJ: Lawrence Erlbaum Associates.

Hoban, G., Nielsen, W., & Hyland, C. (2016). Blended media: Student-generated mash-ups to promote engagement with science content. *International Journal of Mobile and Blended Learning, 8*(3), 35–48.

Jewitt, C. (Ed.). (2014). *Routledge handbook of multimodal analysis*. London: Routledge.

Kress, G. (2010). *Multimodality: A social semiotic approach to contemporary communication*. London: Routledge.

Loughran, J., Mulhall, P., & Berry, A. (2008). Exploring pedagogical content knowledge in science teacher education: A case study. *International Journal of Science Education, 30*, 1301–1320.

New South Wales Board of Studies [NSW BOS]. (1991). *Science and technology K-6 syllabus and support document*. Sydney: New South Wales Board of Studies.

New South Wales Board of Studies [NSW BOS]. (2002). *Mathematics K-6: Syllabus 2002*. Sydney: New South Wales Board of Studies.

New South Wales Board of Studies [NSW BOS]. (2012a). NSW *syllabus mathematics K-10* (Vol. 2). Sydney: New South Wales Board of Studies.

New South Wales Board of Studies [NSW BOS]. (2012b). *Science K-10 (Incorporating science and technology K-6) syllabus*. Sydney: New South Wales Board of Studies.

New South Wales Board of Studies [NSW BOS]. (2016). *A guide to coding and computational thinking across the curriculum*. Retrieved November 9, 2017, from http://educationstandards.nsw.edu.au/wps/portal/nesa/k-10/learning-areas/technologies/coding-across-the-curriculum

New South Wales Department of Education and Training [NSW DET]. (2003). *Quality teaching in* NSW *public schools: A discussion paper*. Ryde: New South Wales Department of Education and Training.

New South Wales Department of School Education. (1989). *Mathematics K–6*. Sydney: New South Wales Department of School Education.

New South Wales Education Standards Authority [NESA]. (2017a). *Program accreditation requirements*. Retrieved from http://educationstandards.nsw.edu.au

New South Wales Education Standards Authority [NESA]. (2017b). *Science and technology K-6*. Sydney: New South Wales Education Standards Authority.

New South Wales Education Standards Authority [NESA]. (2017c). TAS (*Technological and Applied Studies*). Sydney: New South Wales Education Standards Authority.

Nielsen, W., Hoban, G., & Hyland, C. (2017). Pharmacology students' perceptions of creating multimodal digital explanations. *Chemistry Education Research and Practice, 18*, 329–339.

Schank, R. C., Berman, T. R., & Macpherson, K. A. (1999). Learning by doing. In C. M. Reigeluth (Ed.), *Instructional-design theories and models: A new paradigm of instructional theory* (Vol. 2, pp. 161–182). Mahwah, NJ: Routledge.

Skemp, R. R. (1976). Relational understanding and instrumental understanding. *Mathematics Teaching, 77*, 20–26.

Tytler, R., Prain, V., Hubber, P., & Waldrip, B. (Eds.). (2014). *Constructing representations to learn in science*. Dordrecht: Sense Publishers.

Zembal-Saul, C., McNeill, K. L., & Hershberger, K. (2013). *What's your evidence?* Boston, MA: Pearson.

CHAPTER 10

Primary Pre-Service Teachers' Perceptions of STEM Education: Conceptualisations and Psychosocial Factors

Grant Cooper and Nicky Carr

Abstract

The aim of this paper is to examine pre-service teachers' (PSTS) perceptions of STEM education, including their conceptualisations and psychosocial factors associated with teaching it. Methods used to in this study to elicit PSTS' perceptions included surveys, online responses and drawings. PSTS in this sample commonly conceptualised STEM education as involving an integrated approach, placing an emphasis on the relationships between disciplines. PSTS also frequently discussed the importance of developing students' generic skills, using problem-based learning and inquiry-related pedagogies. Some participants positioned STEM education as a way of promoting workforce skills and dispositions in their future students. PSTS generally reported positive attitudes to teaching STEM education. They also reported a number of normative influences to teach STEM, however there appeared to be limited opportunities to develop their teaching capacity on professional experience in schools. Relatively low levels of self-efficacy to teach particular areas of STEM were reported by PSTS, particularly engineering and digital technologies. This paper contributes to debates on calls for reform to teacher education programs and discourses about PSTS perceptions of STEM education.

Keywords

STEM education – PSTs and STEM education – PSTs' conceptulisations of STEM education – PSTs' self-efficacy associated with STEM education – psychosocial factors

1 Background

In what has been described as a global policy juggernaut (Carter, 2017), boosting citizens' STEM-related competencies has been viewed by many post-industrial

 | DOI:10.1163/9789004391413_011

nations as crucial to remaining globally competitive, increasing productivity, promoting innovation and maintaining standards of living (Office of the Chief Scientist, 2013; UNESCO, 2017). As a subset of the broader STEM label, STEM education is a contested term with a myriad of (sometimes conflicting) definitions. There are, however, commonalities in the literature. Radloff and Guzey (2016), for instance, identified three common features in the research. First, there is an emphasis on the interconnected nature of disciplines, whereby the level of STEM integration ranges on a continuum. At one extreme, there is a discrete disciplinary approach (akin to S-T-E-M) moving towards a more integrated, trans-disciplinary approach at the opposite extreme (STEM). For STEM education to be genuinely trans-disciplinary, Vasquez (2015) argues that students must undertake real world-problems, contribute to helping shape learning experiences and apply knowledge and skills from different STEM disciplines. Second, connections between school, community and industry are emphasised and include a focus on a range of generic competencies (e.g. problem solving, working in teams, creativity). Last, STEM education is commonly tailored to stakeholders' needs and context. While there are commonalities in the literature, others think that STEM education has an "identity crisis" (p. 2), noting that stakeholder definitions have implications for how it is delivered in schools (Portz, 2015).

A key driver of the motivation of the increased attention on STEM education, at least for some stakeholders, relates to concerns about future work capacity. Commonly, reports about STEM education are underpinned by neoliberal agendas for human capital production, at the expense of other reasons why STEM education is important (Carter, 2017; Hoeg & Bencze, 2017). Workers in STEM fields play a direct role in driving economic growth, with estimates that changing 1% of Australia's workforce into STEM-related roles would add $57.4 billion to Gross Domestic Product (GDP), a measure of international competitiveness (PWC, 2015). Similar findings are also discussed in the US (Donovan, Moreno Mateos, Osborne, & Bisaccio, 2014; Rothwell, 2013) and the UK (Parliament UK, 2017), particularly emphasising the economic and labour market implications of not adequately addressing concerns about a depleted STEM pipeline. Part of the STEM education agenda includes a particular focus on generic skills (e.g. problem solving, digital literacy, presentation skills), sometimes labelled 'enterprise skills' (Foundation for Young Australians, 2016) or 21st century skills (Rotherham & Willingham, 2010). Similarly, the neoliberal themes are palpable, as associations are reported between enterprise skills and higher wages, figures suggesting that demand in job ads for such skills is on the rise (Foundation for Young Australians, 2016). Concern about future work capacity is a salient driver for some in relation to their discussions about the importance of STEM education.

For in-service educators, who themselves may be forming their own conceptualisations about STEM education, Hobbs, Clark, and Plant (2018) stated that

> the challenge for educators is to translate an ill-defined, politically charged and narrowly utilitarian policy agenda of securing a future workforce, into a valid and coherent curriculum ... that addresses the subtle and complex challenge of preparing "twenty-first-century" citizens within the constraints of a traditional school system ... what is needed is a vision that is inclusive and interdisciplinary in nature and specific to school needs. (p. 134)

As in-service educators deal with the challenges of implementing a STEM program in their schools, another important stakeholder to consider is pre-service teachers (PST/PSTs). Future teachers, particularly those with negative views of their own abilities in STEM-related subjects, often articulate a desire to teach children in more effective and engaging ways then they themselves were taught (Cooper & Gilbert, 2016).

The aim of this chapter is to examine PSTs' perceptions of STEM education. How teachers conceptualise, interpret, and subsequently enact STEM education impacts the learning experiences they provide in their classrooms (Diefes-Dux, 2014). Therefore, the authors of this study are interested in PSTs' conceptualisations of STEM education. Additionally, psychosocial factors such as attitude, subjective norm and self-efficacy are reported to be salient predictors of intentions and future behaviours (Ajzen, 2005), including STEM teaching-related contexts (Zint, 2002). Hence, the authors were also interested in eliciting psychosocial measures associated with teaching STEM.

While some research has already been conducted on PSTs' perceptions of STEM education in other countries (Güler, Çakıroğlu, & Yılmaz-Tüzün, 2017; Radloff & Guzey, 2016), there are several distinct features about this study that both highlight its significance and capacity to contribute to debates about STEM education. The authors are unaware of any attempt to examine PSTs' perceptions of STEM from a psychosocial perspective, a lens that offers insights into participants' intended future teaching practices (Cooper, Kenny, & Fraser, 2012; Cooper, Barkatsas, & Strathdee, 2016; Zint, 2002). Moreover, it is important to recognise differences in Australian education systems that may differ from other countries, including for instance, compulsion versus choice in senior secondary STEM subject selection, bifurcation of STEM/non-STEM tracks and different national STEM policies (Marginson, Tytler, Freeman, & Roberts, 2013). Such differences have the potential to impact Australian PSTs' STEM-related experiences in their own schooling and therefore, it could be argued that these differences have the potential to impact their perceptions of STEM education.

The paper is structured into four main sections. First, the authors explain the rationale for eliciting PSTs' *conceptualisations of STEM education* and *psychosocial factors* associated with teaching it. Second, we outline and justify the *method* used in this paper. Third, we present the *results* of the analysis. Last, there is a *discussion* about the implications of this study, research limitations and future research directions.

2 Conceptualisations of STEM Education

Teachers' conceptualisations form a foundational platform for their pedagogy and the subject matter they are teaching, a mental framework implicitly guiding what is of value and what is not, and what concepts are prioritised over others. To Radloff and Guzey (2016), eliciting PSTs' conceptualisations about STEM education is important because first, it furthers knowledge of how teachers' beliefs are aligned with reform efforts of STEM education. Second, it may be helpful in facilitating a discourse about their current conceptualisations and the possible implications of those beliefs on their future pedagogy as a STEM educator. Therefore, the aim was to explore preservice teachers' conceptualisation of STEM education. Additionally, a number of psychosocial measures are elicited in this study, defined here as PSTs' attitudes, subjective norms regarding expectations and self-efficacy to teach STEM. These variables are discussed further below.

3 Psychosocial Factors

3.1 *Attitudes to Teaching STEM*

Attitudes can be viewed as a construct that influences the intensity and direction of behaviour (Ajzen, 2005). Guided by a social cognitive explanation of behaviour, attitude is regarded as a salient predictor of an agent's intentions and future behaviour (Kraft, Rise, Sutton, & Røysamb, 2005), which has been used in the past to understanding teachers' intentions and STEM-related teaching (Zint, 2002). Considering the former, the authors of this paper were particularly interested in PSTs' attitudes regarding the teaching of STEM. Along with his/her conceptualisations, PSTs' attitudes about teaching STEM education should provide further insight into their perceptions about teaching it in schools.

3.2 *Subjective Norms Teaching* STEM

Behaviours are shaped strongly by the social context in which one lives (Fishbein & Cappella, 2006; Cooper, Barkatsas, & Strathdee, 2016). Subjective norms are determined by the person's beliefs about whether important others would approve or disapprove (Ajzen, 2005). In the context of this research, the authors were interested in eliciting PSTs' perceptions regarding normative influences to teach STEM in schools. Possible normative influences considered in this study were university-based teacher educators and staff on professional experience.

3.3 *Self-Efficacy to Teaching* STEM

Self-efficacy can be defined as an individual's' perception that they are sufficiently knowledgeable, skilful, disciplined, and able to perform a particular behaviour (Kraft, Rise, Sutton, & Røysamb, 2005). Self-efficacy beliefs determine how people feel, think, motivate themselves and behave (Bandura, 1997). Low levels of teacher self-efficacy in teaching STEM-related subjects is linked to poor teaching practices including teacher-centred pedagogies, poor questioning and avoidance of teaching concepts considered too difficult (Cooper, Kenny, & Fraser, 2012). In this study, participants' self-efficacy to teach STEM was elicited.

4 Method and Participants

All PSTs at an Australian metropolitan university completing an introductory STEM education elective course were invited to participate in this study. In total, 24 PSTs, or 96% of the students in the STEM education course, agreed to participate in the research. All participants completed surveys, provided responses in Padlet and drew their visualisations. 83% (n=20) identified as female and 16% (n=4) as males. 83% of the sample (n=20) were fourth year students (in a four-year Bachelor of Education degree), while 16% (n=4) third year. All participants reported that they were local domestic students, an age range of between 21–41 years with a mean age of 24. Students reported that they were undertaking the course either because they had a particular interest in STEM or they enjoyed STEM, or they wanted to develop their capacity to teach STEM.

In the context of this study, three data sources were selected. These include surveys administered to participants at the beginning of an elective introduction to STEM education course. A copy of the survey is shown in Appendix A. The survey comprised of four elements including participants' *demographic information* (Section 1), *attitudes* to teaching STEM (Section 2), *normative influences* to teach STEM (Section 3) and *self-efficacy* to teach STEM (Section 4). All Likert questions in the survey were on a seven-point scale. Survey instrument questions for Sections 2, 3 and 4 were adapted from previous studies that have elicited participants' attitudes (e.g. *I like teaching STEM?*), normative influences (e.g. *People who are important to me want me to teach STEM in primary schools?*) and self-efficacy (e.g. *I am confident I can carry out teaching and learning activities that integrate science, technology, engineering and mathematics?*) (Ajzen, 2005; Cooper, Barkatsas, & Strathdee, 2016). This research was approved by the RMIT Human Research Ethics Committee (Reference: CHEAN A 21066-08/17).

Open-ended responses in the form of qualitative data were elicited via students' responses in Padlet. Padlet is a web-application which enables the creation of virtual walls where students and teachers can pose questions, respond and pin up images or files (Deni & Zainal, 2015). Students were asked *What is STEM education*? *How do you define it?* in the first online lecture of the course, allowing them to respond on virtual sticky notes anonymously, allowing others to view and add their response to the question posed. While the possibility to add multimedia was available, participants in this study only responded in text.

Participants were also asked to 'visualise' their confidence to teach STEM. Visualisations can be used to show unseen or intangible phenomena that cannot be directly detected or experienced (Buckley, 2000). They were given examples of possible representations, for instance, an uppercase S to indicate their confidence to teach science concepts, an emoji or some other meaning-making sign associated with each discipline area in STEM. Conversely, if students did not feel confident to teach particular elements of STEM, they might represent this with a lowercase letter or unhappy emojis, something that communicated their negative perceptions on the matter. It was necessary during this process to communicate to students that the researchers were not assuming a discrete disciplinary approach to STEM education, but rather that the visualisations might be an effective tool for evaluating elements of STEM where they may feel different levels of teaching-confidence. The three data sources provided the researchers with an opportunity to validate the results through triangulation (Zeichner & Noffke, 2001).

5 Results

5.1 *Conceptualisations of STEM Education*

Participants reported various conceptualisations of STEM education on the Padlets. Several key themes were noted by the researchers, including an emphasis on discipline integration, generic skills, inquiry-related pedagogies, and highlighting the job-ready associations:

5.1.1 Discipline Integration

A common theme in participants' conceptualisations of STEM education was the emphasis placed on discipline integration. Most participants (95%, *n*=23) mentioned references to discipline integration in their conceptualisations of STEM education, using terms like integrated, interdisciplinary, interlinked, transdisciplinary, combined, interrelated, as typified in the following:

> Science, Technology, Engineering and Mathematics. The theory is that these subjects can be mixed and matched, combined, and/or supplementary to other subject areas, with the aim of a deeper and more effective interdisciplinary approach to education. (Padlet8)

> STEM is a problem based/applied/interdisciplinary learning approach where students apply science, technology, engineering and mathematics in different contexts. (Padlet11)

> Rather than teach the four disciplines as separate and discrete subjects, STEM integrates and bridges them into a cohesive learning paradigm based on real-world applications. (Padlet19)

Two of the preservice teachers took the notion of integration further, identifying STEM education as '*a way of thinking*' (Padlet4), an approach that '*integrates and bridges them* [subjects] *into a cohesive learning paradigm*' (Padlet19).

5.1.2 Generic Skills

The theme of skills was evident in 41% (*n*=10) of the Padlet posts, with statements such as 'STEM is being integrated into schools to develop the skills of learners as we move into the 21st century' (Padlet13) and STEM is 'great for helping students develop 21st century skills' (Padlet18) typical of the response. While it might be fair to say there was a sense of vagueness associated with some of these responses, others elaborated on their definition of 21st century skills by stating:

> STEM learning requires critical thinking. (Padlet19),

> STEM aims to foster inquiring minds, logical reasoning and collaboration skills. (Padlet17)

> The aim of the STEM subject is to allow students to problem-solve. (Padlet16)

> Skills developed using STEM include collaboration, problem solving, creativity. (Padlet13)

5.1.3 Problem-Based Learning (PBL)

Discussion about the advantages of pedagogies aligned with PBL was another prominent feature of participants' responses, with 41% (n=10) mentioning it in their conceptualisation of STEM education. For instance:

> … (students) need to identify a problem. They then create possible solutions to that problem, which can be tested. (Padlet4)

> The aim of the STEM subject is to allow students to problem-solve using the framework of the four subjects, as would a person working in a STEM career would, cutting out the need for any student asking when or why will I ever need to use this. (Padlet16)

When mentioning the perceived advantages of a PBL pedagogy, four preservice teachers made links between PBL and the potential to explore 'real-world' and 'authentic' problems with students. One preservice teacher made a strong claim that STEM education will:

> allow future generations to create more sustainable ways of living, "greener" ways of building homes/buildings, cleaner ways of using transport. Providing a road for todays [sic] learners to be passionate and creative about making critical changes in a fast-paced society. (Padlet6)

Two preservice teachers saw the use of authentic contexts for problem-solving STEM activities as 'more relevant to students' and 'making learning contextual' which in their views made learning more engaging. One preservice teacher reported that:

> STEM lessons are very much engaging as they demonstrate the students why they are learning a certain thing. Knowing the purpose always helps to capture the knowledge more effectively. (Padlet10)

Two participants emphasised the role that authenticity in STEM education could play in 'igniting curiosity.' For instance:

> The value I see in STEM education is that it activates curiosity and aids people in 'decoding' things that we use/are immersed in every day. (Padlet5)

Another theme that emerged from the Padlet posts about STEM pedagogy was the importance of experiences involving hands-on learning, where theory and conceptual understandings are applied towards solving authentic problems.

5.1.4 Inquiry-Related Pedagogies

Participants associated inquiry-based pedagogies with their conceptualisations of STEM education, with 29% (n=7) of the Padlet posts explicitly mentioning inquiry as an approach. Phrases such as '*exploring through inquiry*' (Padlet23) and '*inquiry-based learning*' (Padlet1 & 3) were reported. One preservice teacher provided this conceptualisation of STEM education:

> Students are able to make connections with theory and practice by linking concepts to the real world that require risk-taking challenges that are hands-on, experimental, collaborative, inclusive and creative that develop skills in problem-solving and inquiry. (Padlet22)

5.1.5 Preparation for Work

Five PSTs (21%) related future job prospects with their conceptualisations of STEM education. As exemplified in the following posts, it appears as if some of the work-focused rhetoric underpins PSTs' conceptualisations of STEM education:

> The emphasis placed on STEM in the modern classroom is becoming increasingly popular with more demand growing for workers in STEM-like fields. (Padlet21)

> STEM education is about preparing future learners for the growing world and the creation of new jobs to meet the needs of the changing and developing world. (Padlet5)

5.2 *Attitudes*

As shown in Figure 10.1, there appears to be a largely positive attitude to teaching STEM in the sample. Students reported a relatively high level of fun when teaching STEM ($\bar{x}$=5.92), liking of teaching it ($\bar{x}$=5.25) and enjoyment ($\bar{x}$=5.25).

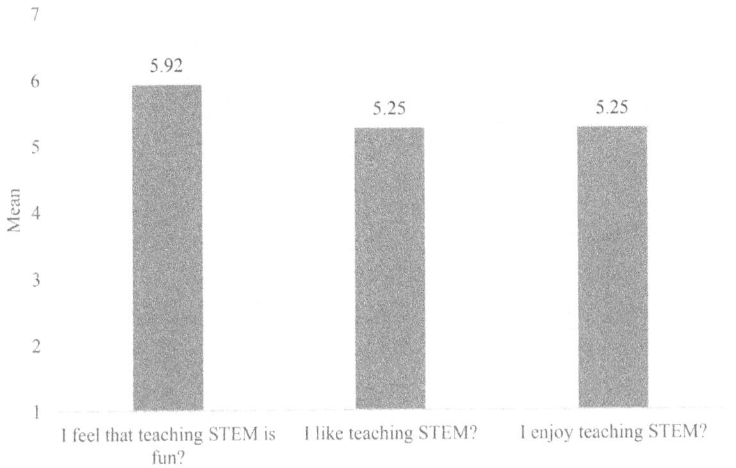

FIGURE 10.1 Attitudes to teaching STEM

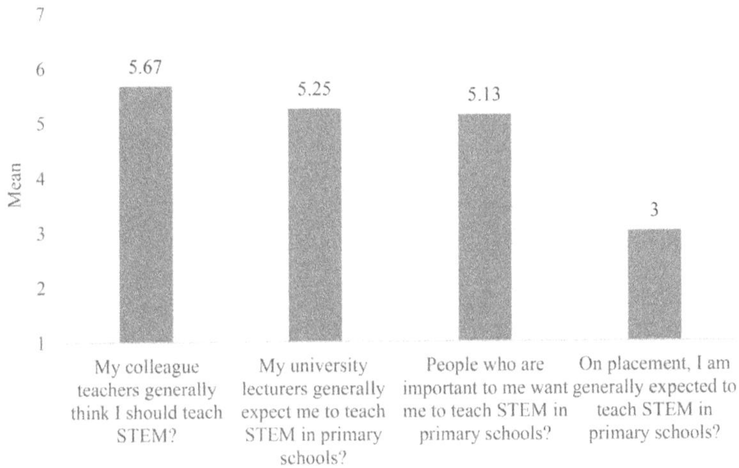

FIGURE 10.2 Subjective norm to teach STEM

5.3 *Subjective Norm to Teach* STEM

Shown in Figure 10.2, PSTs reported relatively strong subjective norms to teach STEM ($\bar{x}$=5.13). In relation to specific stakeholders, participants reported colleague teachers as their strongest normative influence to teach STEM ($\bar{x}$=5.67). In the survey, colleague teacher was the term used to refer to the supervising teachers PSTs would work with on professional experience in primary classroom. However, it was interesting to note that while PSTs felt that their colleague teacher thought they should teach STEM at some stage in their teacher journey, there was on average a relatively low level of expectation to teach it on professional experience ($\bar{x}$=3).

5.4 *Self-Efficacy*

Figure 10.3 shows the STEM self-efficacy measures elicited from participants. PSTs' responses were similar scores on all four questions ($\bar{x}$=4.87, 4.67, 4.67, 4.08). Participants' overall self-efficacy mean score ($\bar{x}$=4.57) was lower than their overall mean attitudinal responses ($\bar{x}$=5.47) and reported normative influences ($\bar{x}$=4.76). To examine if PSTs were more or less confident with particular elements of teaching STEM, they were asked to 'visualise' their confidence to teach each STEM-related discipline.

Two examples of participants' visualisations are shown in Figure 10.4. PSTs' drawings were analysed, using two forms of criteria including first, upper/lower case letters and second, emojis in the form of faces expressing smiles or frowns. From these drawings, the researchers evaluated if the participants' drawing indicated confidence to teach different STEM-related disciplines (e.g. uppercase letter/happy face) or not (e.g. lower-case letter/unhappy face).

As shown in Table 10.1, PSTs indicated different levels of confidence across the STEM disciplines. A large majority of PSTs indicated confidence to teach maths (83%, n=20) while nearly three quarters of the sample reported confidence to teach science (71%, n=17). Conversely, 91% (n=22) indicated that they were not confident to teach engineering-related concepts. Approximately one third of the sample indicated that they were not confident to teach concepts associated with digital technologies (37%, n=9).

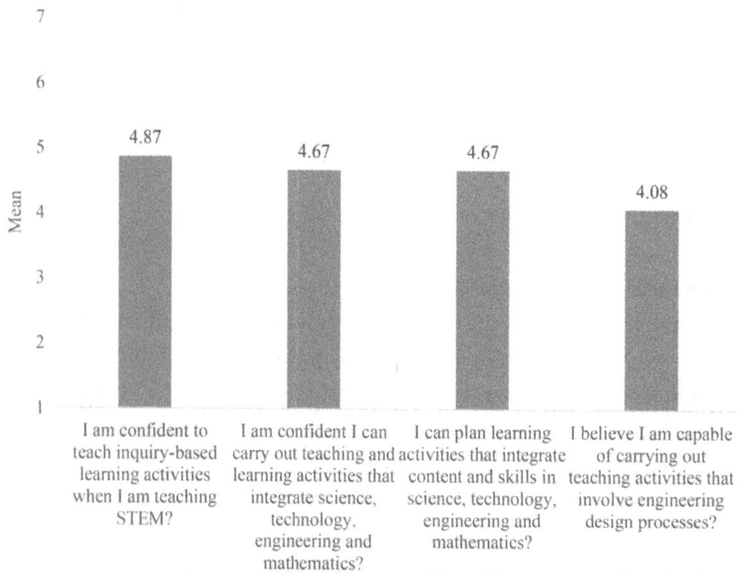

FIGURE 10.3 Self-efficacy measures teach STEM

FIGURE 10.4 Example visualisations

TABLE 10.1 Confidence analysis of participants' drawings

	Confident (n/%)	Not confident (n/%)
Science	17 (71%)	7 (29%)
Digital technologies	15 (63%)	9 (37%)
Engineering-related concepts	2 (9%)	22 (91%)
Maths	20 (83%)	4 (17%)

6 Discussion

6.1 *PSTs' Conceptualisations of STEM Education*

The results of this study suggest that PSTs conceptualised STEM education in a number of ways and comprising various elements. The most dominant theme in participants' conceptualisations of STEM education was the emphasis on the relationships between disciplines, emphasising discipline integration. Similar findings were reported in Radloff and Guzey (2016), who found that PSTs in their study commonly emphasised the integrated or interconnected nature of STEM education. It is perhaps likely that terms such integrated, interdisciplinary, transdisciplinary and interrelated were used interchangeably by participants, perhaps without ample awareness that these terms carry different definitions. Discussions and activities with PSTs that highlight the subtle but important differences between such terms may be an effective pedagogical tool to facilitate students' conceptualisations. Such experiences may be helpful in exploring various models of teaching STEM education with PSTs, viewing it somewhere on a continuum, ranging from a discrete disciplinary approach to a more integrated trans-disciplinary approach (Vasquez, 2015).

There was an emphasis on the instructional elements in participants' STEM conceptualisations. Examples included, for instance, problem-based learning, authentic learning and inquiry-based pedagogies. Consistent with similar research internationally, PSTs' conceptualisations of STEM education had an instructional focus, emphasising connections between school, community and industry across real-life contexts (Radloff & Guzey, 2016). A focus on problem-based, authentic learning aligns more with a transdisciplinary integration of STEM education where the four disciplines are integrated into "real-world, rigorous, relevant learning experiences for students" (Vasquez, 2015, p. 11). There appeared to be a particular emphasis on an instructional-focus of STEM education, with a paucity of responses that addressed content knowledge within STEM – the PSTs in this study did not mention content knowledge related to disciplines that make up STEM education, instead placing more emphasis on conceptualising STEM as a vehicle for the development of skills.

Participants emphasised generic skills in their conceptualisations of STEM education. It was apparent that some PSTs adopted the '21st-century skill' rhetoric, perhaps failing to acknowledge that critical thinking and problem solving, for instance, have been components of human progress throughout history and are not exclusive to this century (Rotherham & Willingham, 2010). Participants emphasised their future students' capacities to critically think, collaborate and problem solve as an important element in their STEM conceptualisations. Similar ideas are discussed in a report by the Foundation for Young Australians (FYA, 2016), called: *The new basics: Big data reveals the skills young people need for the New Work Order*. In this report, enterprise skills including problem solving, financial literacy, critical thinking, creativity, teamwork, digital literacy and presentation skills are discussed, including estimates about future demand for enterprise skills in jobs of the future (FYA, 2016). While such skills are important for students as they navigate a complex future (including in their future workplaces), the discourse about generic skills is overly dominated by neoliberal themes. PSTs perhaps need to have a concurrent awareness that the development of generic skills is likely to help their future students solve the challenges that lie ahead and are therefore important to promote *and* that dominant themes about human capital production underpin the generic skills literature, with a focus on securing a future workforce. PSTs need to be effectively prepared to read such texts with the capacity to identify author bias. Approximately one fifth of the sample explicitly associated their future students' job prospects with their conceptualisations of STEM education. This is hardly surprising as much of the STEM education literature is underpinned by neoliberal biases (Carter, 2017). Discussions and activities with PSTs that explore various rationales for why they should promote and foster generic skills with their future students are important, perhaps as part of a broader evaluation examining opportunities, limitations and barriers to teaching STEM education.

6.2 *Attitudes to Teaching STEM*

PSTs in this study reported generally positive attitudes to teaching STEM. Their positive attitudes were exemplified by their reported enjoyment, fun and liking of teaching STEM. Attitudes are commonly a salient predictor of intentions and future behaviour (Ajzen, 2005), and may be indicative of PSTs' broader perceptions about particular subject disciplines and their propensity to teach it effectively (Zint, 2002). This research suggests that PSTs should have opportunities in their university course to consider possible reasons for their positive or negative attitudes to teaching STEM. One potential reason for PSTs' negative attitudes to teaching STEM may be associated with their own experiences in school. Future teachers, particularly those with negative views of their own abilities in STEM-related disciplines, often articulate a desire to teach children in more effective and engaging ways than they themselves were taught (Cooper & Gilbert, 2016). An important initial step in changing one's attitude is firstly unpacking the behavioural beliefs that form it (Ajzen, 2005), potentially increasing PST awareness on why they may feel a particular way towards teaching STEM-related content.

6.3 *Subjective Norms to Teach STEM*

Overall, PSTs reported a relatively high degree of normative influence to teach STEM from stakeholders such as colleague teachers and university-teacher educators. It was interesting to note that while PSTs typically perceived a relatively strong influence to teach STEM from colleague teachers, they concurrently reported relatively low expectations to teach STEM on professional experience in schools. Research indicates that many in-service teachers generally have narrow content knowledge in STEM areas and do not have sufficient instructional resources, pedagogical knowledge or teaching approaches that enable them to support STEM learning (Chiu, Price, & Ovrahim, 2015). Hence, one possibility may be that while in-service teachers generally view the idea of STEM education in a favourable way, and this positive evaluation is communicated to PSTs in some form, perceptions about their own capacity to teach STEM may translate into avoidance. Such a disposition may manifest into low expectations on PSTs to teach it whilst on professional experience. Similar outcomes are mentioned in the science education literature; with research to indicate that many in-service teachers lack content knowledge and confidence to teach science (Lawrence & Palmer, 2003). This perpetuates the under-teaching of science in primary schools because many PSTs have limited opportunities to observe experienced teachers modelling effective science teaching or get an opportunity to actually teach science themselves during their professional experience (Kenny, 2010). Similar findings are noted elsewhere in other STEM disciplines, including digital technologies (Ertmer & Ottenbriet-Leftwish, 2010; Prestridge, 2012) and engineering (Hammock & Ivey, 2017). The reported lack of STEM experiences in schools may highlight the need to include more STEM experiences in the university component of their teaching degrees, including experiences that university educators can provide, but also through additional STEM-focused school-based experiences. There may be opportunities to combine course/unit subjects in ways that emphasise the interdisciplinary nature of STEM education, overcoming the traditional siloed approach that typically pervades the structure of many courses. Similar recommendations have been made internationally (Radloff & Guzey, 2016), emphasising the value of modelling effective discipline integration and STEM-related pedagogies in teaching courses (e.g. PBL, real world contexts, inquiry approaches).

6.4 *STEM Teaching Self-Efficacy*

An important element of this study was to measure PSTs' self-efficacy to teach STEM. The scores in the self-efficacy section of the survey were relatively lower in comparison to participants' attitudes and subjective norms. This result may be indicative of the relatively low level of confidence PSTs typically have in relation to teaching STEM. As discussed, low levels of teacher self-efficacy in teaching STEM-related disciplines is linked to poor teaching practices such as teacher-centred pedagogy, poor questioning and avoidance of teaching concepts considered too difficult (Cooper, Kenny, & Fraser, 2012). Early career teachers often revert back to safe or familiar teaching methods: either what they experienced growing up, or similar to the teaching styles of their colleague teachers (Rodriguez, 2015). Low levels of self-efficacy and a lack of STEM experiences in their teacher preparation courses runs the risk that PSTs generally will not embrace the STEM education movement, instead reverting to safe or familiar ways of teaching, approaches that may not align with a transdisciplinary approach to teaching STEM.

The visualisations presented a more concerning response, in which an overwhelming majority of PSTs reported not feeling confident to teach engineering-related concepts. English (2016) argued that the STEM education label has mainly been used in reference to science, with less emphasis on the remaining disciplines – especially engineering. In an Australian context particularly "one aspect that remains in need of greater attention is the relative lack of inclusion of engineering experiences in STEM curricula, especially in the primary grades" (English, 2016, p. 85). With such a large proportion of PSTs in the study reporting not feeling confident to teach engineering-related concepts, there appears to be a need for stakeholders to consider the amount and types of related experiences. It is also important for PSTs to see increases in the amount of engineering-related thinking and designing being taught while on professional experience.

A relatively large proportion of the PSTs in this study reported not feeling confident to teach digital technologies. Similar to the calls to improve the teaching of engineering-related concepts, PSTs need to experience innovative and authentic ways of learning and teaching through digital technologies in their teaching course (Carr, 2016). Conversely, the majority of PSTs in this sample reported confidence to teach maths and science. It is unclear from the visualisation if PSTs' confidence is associated with their discipline-related knowledge or capacity to teach it effectively. It is important to note that the visualisations elicited in this research are an exploratory research method. Such methods appear to be an effective way for participants to represent their confidence to teach particular disciplines. One possible limitation of this technique, however, is that it may classify confidence into binary categories (e.g. confident/not confident); further refinement of this approach, for example if combined with interviews or focus groups, may be able to elicit more accuracy of PSTs' confidence to teach particular subjects.

Overall, the results indicate that many PSTs do not feel properly prepared to teach STEM. It is important to keep in mind that participants in this study were all completing an elective STEM education course.

Therefore, PSTs with a particular interest, background and confidence in STEM might have been more likely to enrol. Alternatively, participants may have enrolled in the STEM course because of concerns (e.g. poor self-efficacy) about their own capacities. The sample size is relatively small and there is no suggestion that it was representative, but the study does provide valuable insights about PSTs' perceptions of STEM education. Future research may explore PSTs' perceptions of STEM education using a larger, representative sample across multiple universities, examining conceptualisations and factors associated with teaching STEM. Other possible foci may be PSTs' self-efficacy to teach STEM during professional experiences and as early career teachers.

7 Conclusion

The aim of this paper was to examine PSTs' perceptions of STEM education, including their conceptualisations and psychosocial factors associated with teaching it. The findings indicated that PSTs commonly conceptualised STEM education as involving an integrated approach, placing emphasis on the relationships between disciplines. PSTs also frequently discussed the importance of developing students' generic skills, using problem-based learning and inquiry-related pedagogies. Some also positioned STEM education as a vehicle for preparing students for future work. PSTs generally reported positive attitudes to teaching STEM education. PSTs reported a number of normative influences from stakeholders, however there appeared to be generally limited opportunities to develop their STEM teaching capacity on professional experience. PSTs' reported relatively low levels of self-efficacy to teach particular areas of STEM, particularly engineering and digital technologies. The findings suggest that initial teacher educators may need to reform their current approaches to STEM education to:

- help PSTs take a more critical view of how STEM education and so-called 21st century skills are positioned in policy;
- assist PSTs to identify the nuanced terms that describe different models of integrated learning, such as interdisciplinary, transdisciplinary and so on;
- provide authentic opportunities for PSTs to develop their STEM teaching capacity in educational settings, preferably in less siloed ways than traditionally adopted; and
- focus on increasing PSTs' self-efficacy in engineering and digital technologies.

References

Ajzen, I. (2005). *Attitudes, personality, and behavior* (1st ed.). Berkshire: Open University Press.

Bandura, A. (1997). *Self-efficacy*. New York, NY: Academic Press.

Buckley, B. (2000). Interactive multimedia and model-based learning in biology. *International Journal of Science Education, 22*(9), 895–935. Retrieved from http://dx.doi.org/10.1080/095006900416848

Carr, N. (2016). Pre-service teachers teaching about and across cultures using digital environments: The case of eTutor. *Educational Media International, 53*(2), 103–117. Retrieved from http://dx.doi.org/10.1080/09523987.2016.1211336

Carter, L. (2017). Neoliberalism and STEM education: Some Australian policy discourse. *Canadian Journal of Science, Mathematics and Technology Education, 17*(4), 247–257. Retrieved from http://dx.doi.org/10.1080/14926156.2017.1380868

Chiu, A., Price, C. A., & Ovrahim, E. (2015). *Supporting elementary and middle school STEM education at the whole school level: A review of the literature*. Paper presented at the NARST 2015 Annual Conference, Chicago, IL.

Cooper, G., Barkatsas, T., & Strathdee, R. (2016). The Theory of Planned Behaviour (TPB) in educational research using Structural Equation Modelling (SEM). In T. Barkatsas & A. Bertram (Eds.), *Global learning in the 21st century* (pp. 139–162). Rotterdam, The Netherlands: Sense Publishers.

Cooper, G., & Gilbert, A. (2016). Using moments of wonder in science with pre-service teachers. *Asia-Pacific Forum on Science Learning and Teaching, 17*(2), 1–27.

Cooper, G., Kenny, J., & Fraser, S. (2012). Influencing intended teaching practice: Exploring pre-service teachers' perceptions of science teaching resources. *International Journal of Science Education, 34*(12), 1883–1908. Retrieved from http://dx.doi.org/10.1080/09500693.2012.698762

Deni, A., & Zainal, Z. (2015). Let's write on the wall: Virtual collaborative learning using padlet. *Turkish Online Journal of Educational Technology, 2*, 364–369.

Diefes-Dux, H. A. (2014). In-service teacher professional development in engineering education: Early years. In S. Purzer, J. Strobel, & M. Cardella (Eds.), *Engineering in precollege settings: Synthesizing research, policy, and practices*. Lafayette: Purdue University Press.

Donovan, B., Moreno Mateos, D., Osborne, J., & Bisaccio, D. (2014). Revising the economic imperative for US STEM education. *Plos Biology, 12*(1), e1001760. Retrieved from http://dx.doi.org/10.1371/journal.pbio.1001760

English, L. (2016). *Targeting all of stem in the primary school: Engineering design as a foundational process*. Retrieved February 10, 2018, from https://research.acer.edu.au/cgi/viewcontent.cgi?referer=https://www.google.com.au/&httpsredir=1&article=1298&context=research_conference

Ertmer, P. A., & Ottenbriet-Leftwish, A. T. (2010). Teacher technology change: How knowledge, confidence, beliefs & culture intersect. *Journal of Research on Technology in Education, 42*(3), 255–284. Retrieved from https://doi.org/10.1080/15391523.2010.10782551

Fishbein, M., & Cappella, J. (2006). The role of theory in developing effective health communications. *Journal of Communication, 56*(s1), 1–17. Retrieved from https://doi.org/10.1111/j.1460-2466.2006.00280.x

Güler, F., Çakıroğlu, J., & Yılmaz-Tüzün, O. (2017). *ECER programmes*. Retrieved February 9, 2018, from http://www.eera-ecer.de/ecer-programmes/conference/22/contribution/41771/

Hair, J., Black, W., Babin, B., & Anderson, R. (2014). *Multivariate data analysis* (7th ed.). Upper Saddle River, NJ: Prentice Hall.

Hammock, R., & Ivey, T. (2017). Examining elementary teachers' engineering self-efficacy and engineering teacher efficacy. *School Science and Mathematics, 117*, 52–62. doi:/10.1111/ssm.12205

Hobbs, L., Cripps Clark, J., & Plant, B. (2018). Successful students – stem program: Teacher learning through a multifaceted vision for stem education. In R. Jorgensen & K. Larkin (Eds.), *STEM education in the junior secondary: The state of play*. Singapore: Springer.

Hoeg, D., & Bencze, J. (2017). Values underpinning STEM education in the USA: An analysis of the next generation science standards. *Science Education, 101*(2), 278–301. Retrieved from http://dx.doi.org/10.1002/sce.21260

Johnson, R., & Onwuegbuzie, A. (2004). Mixed methods research: A research paradigm whose time has come. *Educational Researcher, 33*(7), 14–26. Retrieved from http://dx.doi.org/10.3102/0013189x033007014

Kenny, J. (2010). Preparing primary teachers to teach primary science: A partnership based approach. *International Journal of Science Education, 32*(10), 1267–1288. Retrieved from https://doi.org/10.1080/09500690902977994

Kraft, P., Rise, J., Sutton, S., & Røysamb, E. (2005). Perceived difficulty in the theory of planned behaviour: Perceived behavioural control or affective attitude? *British Journal of Social Psychology, 44*(3), 479–496. Retrieved from http://dx.doi.org/10.1348/014466604x17533

Lawrence, G. A., & Palmer, D. H. (2003). *Clever teachers, clever sciences: Preparing teachers for the challenge of teaching science, mathematics and technology in 21st century Australia*. Retrieved February 9, 2018, from http://www.dest.gov.au/sectors/higher_education/publications

Marginson, S., Tytler, R., Freeman, B., & Roberts, K. (2013). *STEM: Country comparisons: International comparisons of Science, Technology, Engineering and Mathematics (STEM) education*. Melbourne: Australian Council of Learned Academies.

Office of the Chief Scientist. (2013). *Science, technology, engineering and mathematics in the national interest*. Canberra: Office of the Chief Scientist.

Parliament UK. (2017). *Industrial strategy: Science and STEM skills.* Retrieved from https://publications.parliament.uk/pa/cm201617/cmselect/cmsctech/991/99106.htm

Prestridge, S. (2012). The beliefs behind the teacher that influences their ICT practices. *Computers & Education, 58*(1), 449–458. Retrieved from https://doi.org/10.1016/j.compedu.2011.08.028

Prinsley, R., & Baranyai, K. (2015). *Cite a website: Cite this for me.* Retrieved February 9, 2018, from http://www.chiefscientist.gov.au/wp-content/uploads/OPS09_02Mar2015_Web.pdf

PWC. (2015). *A smart move: Future-proofing Australia's workforce by growing skills in Science, Technology, Engineering and Maths (STEM).* Retrieved February 9, 2018, from https://www.pwc.com.au/pdf/a-smart-move-pwc-stem-report-april-2015.pdf

Radloff, J., & Guzey, S. (2016). Investigating preservice STEM teacher conceptions of STEM education. *Journal of Science Education and Technology, 25*(5), 759–774. Retrieved from http://dx.doi.org/10.1007/s10956-016-9633-5

Rodriguez, A. (2015). What about a dimension of engagement, equity, and diversity practices? A critique of the next generation science standards. *Journal of Research in Science Teaching, 52*(7), 1031–1051. Retrieved from https://doi.org/10.1002/tea.21232

Rotherham, A., & Willingham, D. (2010). *21st century skills: Not new but a worthy challenge.* Retrieved February 9, 2018, from http://www.aft.org/sites/default/files/periodicals/RotherhamWillingham.pdf

Rothwell, J. (2013). *The hidden STEM economy.* Retrieved February 9, 2018, from https://www.brookings.edu/wp-content/uploads/2016/06/TheHiddenSTEMEconomy610.pdf

Siekmann, G. (2016). *What is STEM? The need for unpacking its definitions and applications.* Adelaide: National Centre for Vocational Education Research (NCVER).

Tashakkori, A., & Teddlie, C. (2010). Putting the human back in "human research methodology": The Researcher in mixed methods research. *Journal of Mixed Methods Research, 4*(4), 271–277. Retrieved from http://dx.doi.org/10.1177/1558689810382532

UNESCO. (2017). *Cracking the code: Girls' and women's education in Science, Technology, Engineering and Mathematics (STEM).* Retrieved February 5, 2018, from http://unesdoc.unesco.org/images/0025/002534/253479E.pdf

Vasquez, J. (2015). *STEM-beyond the acronym.* Retrieved February 9, 2018, from http://www.ascd.org/publications/educational-leadership/dec14/vol72/num04/STEM%E2%80%94Beyond-the-Acronym.aspx

Zeichner, K. M., & Noffke, S. (2001). Practitioner research. In V. Richardson (Ed.), *Handbook of research on teaching* (4th ed., pp. 298–330). Washington, DC: American Educational Research Association.

Zint, M. (2002). Comparing three attitude-behavior theories for predicting science teachers' intentions. *Research in Science Teaching, 39*(9), 819–844. Retrieved from http://dx.doi.org/10.1002/tea.10047

Appendix A: STEM Teaching and Perceptions Scale

Selection 1: Some general information about you

Thankyouforparticipatinginthisstudy.Itisimportanttopointoutthattherearenoright or wrong answers; I'm interested in your beliefs about STEM (Science, Technology, Engineering & Maths) education

1. Are you male, female or other? (Please tick below)
   - Female
   - Male
   - Other
2. What is your age?

   ______
3. Do you identify as any of the following? (Please tick below)
   - Local domestic student
   - International on-shore student
   - Aboriginal or Torres-Strait Islander
4. What year best describes your progress in your degree? (Please tick below)
   - 1st
   - 2nd
   - 3rd
   - 4th

Selection 2: Please indicate your response to the following questions/statements:

I like teaching STEM?	Strongly disagree	1	2	3	4	5	6	7	Strongly agree
I enjoy teaching STEM?	Strongly disagree	1	2	3	4	5	6	7	Strongly agree
I feel that teaching STEM is fun?	Strongly disagree	1	2	3	4	5	6	7	Strongly agree

Selection 3: Please indicate your response to the following questions/ statements:

People who are important to me want me to teach STEM in primary schools?	Strongly disagree	1	2	3	4	5	6	7	Strongly agree
My colleague teachers generally think I should teach STEM?	Strongly disagree	1	2	3	4	5	6	7	Strongly agree
My university lecturers generally expect me to teach STEM in primary schools?	Strongly disagree	1	2	3	4	5	6	7	Strongly agree
On placement, I am generally expected to teach STEM in primary schools	Strongly disagree	1	2	3	4	5	6	7	Strongly agree

Selection 4: Please indicate your response to the following questions/ statements:

I am confident I can carry out teaching and learning activities that integrate science, technology, engineering and mathematics?	Strongly disagree	1	2	3	4	5	6	7	Strongly agree
I can plan learning activities that integrate content and skills in science, technology, engineering and mathematics?	Strongly disagree	1	2	3	4	5	6	7	Strongly agree
I am confident to teach inquiry-based learning activities when I am teaching STEM?	Strongly disagree	1	2	3	4	5	6	7	Strongly agree
I believe I am capable of carrying out teaching activities that involve engineering design processes?	Strongly disagree	1	2	3	4	5	6	7	Strongly agree

CHAPTER 11

Building STEM Self-Perception and Capacity in Pre-Service Science Teachers through a School-University Mentor Program

Amanda Berry, Tricia McLaughlin and Grant Cooper

Abstract

This chapter reports a research project aimed to develop pre-service science teachers' knowledge and understanding of contemporary STEM contexts and pedagogies through participation in a STEM mentoring initiative for schoolgirls. In this project, primary and secondary pre-service teachers (PSTs) volunteered to work as mentors, collaborating in the design of learning experiences suitable for school-aged girls, together with teacher educators and researchers in STEM at an Australian University. Outcomes of the study focus on main themes of: PSTs' self-perceptions as emerging STEM educators, their understandings of STEM and developing a pedagogy around STEM, their understandings of school girls' interest, engagement and learning in STEM, and the value of the project for teachers in preparation.

Keywords

Pre-service science teachers – mentoring – STEM identity – school-university partnership

1 Introduction

Increasing the number of future graduates entering Science, Technologies, Engineering and Mathematics (STEM) careers and developing the STEM skills of all citizens is the focus of considerable political, industry and media attention both in Australia and internationally. For example, in Australia, the importance of a well-qualified and diverse STEM workforce is widely acknowledged as crucial to Australia's long-term economic future (Office of Chief Scientist, 2016). Hence STEM education is a national priority that is seen as both contributing to future workforce needs and shaping the growth of scientifically and

 | DOI:10.1163/9789004391413_012

technologically literate citizens who can critically examine, understand and improve the world around them.

STEM education spans a variety of understandings and translations in the school context, but integral to any understanding is the role of the teacher. Teachers have an important responsibility to effectively support learning in STEM through developing students' knowledge, skills and attitudes in both a foundational and integrated manner that inspires more students to take up future work and learning opportunities in the STEM fields. In particular, engaging the interest and participation of school-aged girls is considered vital, as females have long been under-represented in the STEM disciplines at school, university and business levels, both in Australia and internationally (Office of the Chief Scientist, 2016; UNESCO, 2017).

Yet there is evidence that many teachers both in Australia and internationally do not know what STEM education actually entails; have little understanding of how to develop integrated STEM activities in their classrooms; have narrow content knowledge in STEM areas; and do not have sufficient instructional resources, pedagogical knowledge or teaching approaches that enable them to support STEM learning in their students (Breiner, Johnson, Sheats Harkness, & Koehler, 2012; Chiu, Price, & Ovrahim, 2015). Difficulties associated with STEM understanding and conceptualisation are also evident in pre-service teachers whose teaching efforts often mirror the practices of their own prior experiences of schooling, or of their supervising teachers during practicum, that limits opportunities for growing their STEM capacities and identities (Windschitl, 2004; Kim, Yuan, Hill, Doshi, & Thai, 2015).

Two important elements in promoting change in STEM education and wider future engagement in STEM of all learners are building the capacity of pre-service teachers (both primary and secondary) in the STEM fields and, supporting the development of pre-service teachers' (PSTS') self-conceptions or identity as future STEM educators. This chapter reports research from a project that aimed to develop PSTS' knowledge and understanding of contemporary STEM contexts and pedagogies and build their perceptions of themselves as STEM educators through participation in a STEM mentoring initiative for schoolgirls. The study explored the role of mentoring as a means of supporting PSTS' capacity building and self-perception as emerging STEM educators.

2 Conceptions of STEM

STEM education itself, both within Australia and internationally, is a contested term with a myriad of (sometimes conflicting) definitions, although there appear to be commonalities in the literature. Radloff and Guzey (2016) identified three common features of STEM education as:

- a set of integrated or interconnected disciplines, whereby the level of STEM integration ranges on a continuum, from a discrete disciplinary approach (akin to S-T-E-M) to a more integrated, trans-disciplinary approach (STEM) (Vasquez, 2015);
- an instructional focus, where connections between school, community and industry are emphasised across real-life contexts, encompassing a range of generic competencies (e.g. problem solving, working in teams, creativity) and;
- a tailoring of the focus of STEM to stakeholders' needs and contexts.

These common features, whilst contributing to a greater theoretical understanding of STEM education, do not promote comprehension for teachers of how to develop STEM activities or pedagogical approaches in their classrooms. Teachers in schools, like many others, are also struggling with a clear definition of STEM to apply in their curriculum and teaching approaches.

For in-service educators, who themselves may be forming their own conceptualisations of STEM education, Hobbs, Cripps Clark, and Plant (2017, p. 134) stated,

> the challenge for educators is to translate an ill-defined, politically charged and narrowly utilitarian policy agenda of securing a future workforce, into a valid and coherent curriculum ... that addresses the subtle and complex challenge of preparing "twenty-first-century" citizens within the constraints of a traditional school system.

This is particularly the case for primary teachers who typically bring very limited prior exposure and poor attitudes to STEM fields (Nadelson, Callahan, Pyke, Hay, Dance, & Pfiester, 2013).

3 STEM Identity and Self-Perception

Policy-makers, legislators and educational experts have called for changes in teacher education programmes to develop and promote broader understandings of STEM amongst pre-service teachers (Breiner et al., 2012). However, there is evidence that many PSTs are not implementing STEM educational approaches in their own classrooms upon graduation despite preparatory experiences in their pre-service programmes (Johnson, 2013). While opportunities to implement STEM education by PSTs may be stymied by a number of factors, including a school culture and curriculum that privileges particular approaches to subject disciplines and pedagogies, one emerging issue is that of PSTs' perceptions of their own STEM identities and their perceptions of themselves as STEM educators. Carlone and Johnson (2007) define STEM identity in terms of 'fitting in' within STEM fields, specifically, the way individuals make "meaning of science experiences and how society structures possible meanings" (p. 1187). Polman and Miller (2010) further explain STEM identity as the ability of individuals to see themselves as the kind of people who could be legitimate participants in STEM through their interests, abilities, race, gender, and cultures. Other researchers in identity note that teachers and parents/carers often have different expectations of boys and girls when it comes to STEM and these expectations contribute to an identity that sees STEM as "clever and brainy" and not "caring or nurturing," the latter being traits which are typically seen to be more appealing to girls (Wang & Degol, 2013).

A number of researchers maintain that STEM identity cannot be fully developed unless individuals have opportunities to observe and participate in authentic activities with STEM professionals, often in a mentoring or tutoring relationship (Barab & Hay, 2001; Sadler, 2010). That is, STEM identity can develop progressively through working with expert mentors in authentic contexts. Sjaastad (2012) explored the range of people influencing future STEM choice of school students and distinguished between those acting as *models* – parents, teachers and others displaying a STEM professional identity – and those acting as *definers* – parents or others helping a young person in the process of setting goals, defining values and identifying personal strengths in their educational choices. Building the identity of pre-service teachers as STEM educators was fundamental to this research and the opportunity to participate as STEM mentors was pivotal to this identity building.

4 The Role of Peers and Mentors in the Development of a STEM Identity and Self-Perception

While there are many models of mentoring, mentoring is commonly seen as a practice that involves one person supporting the development of another, usually less experienced, who acts as a role model, guide, counsellor, and/or advocate for the less experienced person (Casey & Shore, 2000). In educational settings, mentoring can have an academic or personal function, or a combination of these, and can be developed through formal programs or informal networks (Wai-Ling Packard & Nguyen, 2003). For example, Heirdsfield, Walker, Walsh, and Wilss (2008) looked at the growth of self-efficacy of mentees in pre-service teacher education, whereby third and fourth year students acted as mentors for first year students. Their research highlighted the value of a relational model of mentoring that can ease transition to new learning or new contexts. The use of peers as mentors and the benefits of engaging in dialogue with peers and supporting one another throughout learning experiences has also been evidenced through Le Cornu's (2007) research. Here, peer mentoring was utilised as a formal strategy for pre-service teachers to become directly involved in each other's learning by acting as mentors for each other.

The role of peers or experts to assist in the development of a STEM identity and self-perception has been identified by a number of researchers (Archer, DeWitt, Osborne, Dillon, Willis, & Wong, 2010; Nadelson et al., 2009). For example, Jaipal-Jarmani and Angeli (2017) identified improvements in PSTs' self-efficacy in STEM through group activities with peers. Stout, Dasgupta, Hunsinger, and McManus (2011) illustrated the benefit of same-sex peer experts in greater self-identification and engagement, which in turn enhanced self-efficacy, domain identification, and commitment to pursue STEM careers in participants. Importantly, these findings identified how the peers' self-concept benefited from contact with STEM experts. Similarly, Tenenbaum, Anderson, Jett, and Yourick (2014) illustrated how an innovative model of mentoring between university students and school students contributed to personal, educational, and professional growth for the mentors and increased the interest and engagement of students studying STEM.

In the research reported in this chapter, primary and secondary PSTs had an opportunity to participate together in a STEM mentoring initiative for schoolgirls, whereby the mentoring was conducted by the participating PSTs for the benefit of each other (peer mentoring) and the school-girls involved.

5 Research Project

The aim of the research project was to investigate the developing self-perceptions of a group of pre-service teachers as STEM educators, through their participation in a university-based STEM mentoring program for schoolgirls aged 12–16 years old. The research project formed part of a larger, nationally funded project designed to promote increased awareness and participation of girls and women in STEM education and to enhance girls' self-confidence and self-identity in STEM (McLaughlin & Berry, 2017). Pre-service teachers from one university, intending to teach in primary and secondary schools, volunteered to work as mentors, collaborating in the design of learning experiences suitable for the schoolgirls.

The research question guiding this study was: *How are pre-service teachers' conceptions of STEM education and their self-perceptions as STEM educators shaped through their experiences of mentoring schoolgirls in a university-school STEM program?*

Figure 11.1 illustrates the intended 'interactional design' framework of the STEM project. Each of the different participant groups and their relationships is represented through the solid and broken lines.

In the framework, some connections are stronger than others and hence are labelled, 'direct.' Other connections are less obvious, labelled in the diagram as 'indirect.' As a key stakeholder in the model, PSTs acted as a *bridge* between other key groups. For instance, PSTs worked closely with schoolgirls, teacher educators and STEM technicians to present STEM experiences in university based workshops. The teacher educators all worked in science education roles in the university teacher education program. The technicians worked in different STEM faculties within the same university, including nanotechnology, advanced materials and manufacturing, robotics and virtual reality. Pre-service teachers concurrently worked with in-service teachers, visiting the girls and teachers in their schools before and after university campus visits, attempting to contextualise and extend the STEM learning purposes and outcomes. Throughout the research project, the PSTs were encouraged to see themselves as STEM educators and mentors of the schoolgirls. The underlying assumption was that the PSTs would interact with the teacher educators and STEM technicians, which would help promote familiarity with and knowledge about STEM

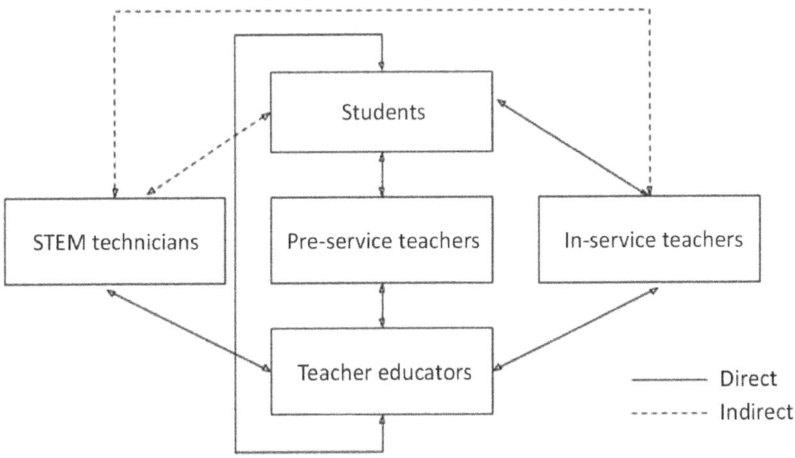

FIGURE 11.1 STEM project mentor framework

contexts and possibilities. The PSTs worked with each other as peer mentors, supporting learning in areas of differing expertise and in designing experiences for the school girls. It was also assumed that working with school girls would further help PSTs understand girls' interests, motivations, knowledge and understanding of STEM. Finally, it was assumed that repeated exposure of PSTs through two iterations of the STEM workshops and accumulating mentoring experiences over an extended period would enhance PSTs knowledge and confidence about STEM and teaching STEM. Information about the workshops offered is detailed in Table 11.1.

The opportunity to take on a STEM mentor role was offered to primary and secondary final year pre-service teachers. It was taken in addition to their regular teacher education program and did not count for course credits. Pre-service teachers were invited to apply for the project, then following selection, 10 PSTs (4 primary and 6 secondary) attended a mentor workshop and met with university-based STEM technicians responsible for designing STEM activities for the schoolgirls. Of the 10 PSTs participating in the mentor program, only 6 provided data for this research – 3 males and 3 females.

Technicians from the four areas mentioned above were expected to work with the PSTs and collaboratively plan activities for the schoolgirls' visits. Pairs of PSTs were then allocated to one of the participating schools and expected to visit the school to meet with the girls who would be attending the STEM

TABLE 11.1 STEM workshops

Workshop	Activities
RMIT Advanced Manufacturing Precinct	Students explored additive and subtractive process manufacturing using a range of materials including 3D printing and scanning.
Nanotech Research Centre	Students visited a nanotech laboratory at the university, exploring research projects across different scientific and technological disciplines including biological, chemistry, physics and biomedical sciences. Students were given the opportunity to consider how such nanotechnology will impact their lives.
Virtual and mixed reality lab	Students experienced the latest VR and AR hardware and software including Microsoft Hololens and Oculus Rift. They also observed industry standard robotics controlled remotely.
Electrical engineering workshop	Students were given the challenge to build an Arduino-operated model car, comprising various elements including infrared sensors.

workshops. PSTs were also expected to provide the school with information about potential links between the university-based STEM activities and the Australian curriculum.

6 Research Approach

The researchers took a socio-cultural perspective on learning by analysing the pre-service teacher-as-mentor experience, as situated in the context of the STEM mentoring project (Lave & Wenger, 1991). Socio-cultural frameworks view learning as a form of participation in social and cultural practices rather than as an internal cognitive process. The situational approach characterizes an individual's development as a social learning experience and process which gradually increases with their desire to participate and their feelings of acceptance and belonging in the situation, locating identity within the community of STEM practice and enactment (Lave & Wenger, 1991). Using a situational socio-cultural model in the context of PST collaborative professional learning and development enabled the PSTs to develop their own STEM educator capacities and identity as they simultaneously developed their role as both mentors for the school girls and each other.

Individual semi-structured interviews were conducted with the participating PSTs before and after their participation in the STEM project. Interviews were recorded and transcribed in full, then analysed according to themes related to the research question. Additional emerging themes were also noted and investigated. University ethics approval to collect data from the pre-service teachers was obtained prior to commencement of data collection.

7 Findings and Discussion

We present our findings according to four main themes derived from the research question: PSTs' conceptions of STEM, their self-perceptions as STEM educators, the building of a STEM identity through mentoring and relationships between purpose and participation. We consider each of these components in relation to PSTs reported experiences before and after the STEM mentoring project.

(i) *Conceptions of STEM*

At the commencement of the project, most of the mentors held broad definitional understandings of STEM related to its specific disciplinary fields, either separately or in combination, and particular kinds of skills. For example,

> STEM is a collection of [disciplines] and scientific thinking critical thinking all kind of rolled in to one. (P1)

> a combination of Science, Technology and Engineering, and Maths … [that in teaching] should be more interlinked and combined. (P5)

Two of the mentors offered ideas about the purposes of STEM

> to tackle really big sort of real world problems. (P2);

and

> as a different way of looking … at data and processes. (P3)

Following the project, mentors showed some small changes in how they talked about their conceptions of STEM and the influence of the mentor program on their ideas.

For example, P1 explained:

> Perhaps I did not articulate it as well as I could have last time. But, I do see STEM more about the skills than it is about the theory involved. So, you are not going to remember specific details, but you are going to remember, or you are going to hopefully have the skills to be able to think scientifically. Be sceptical about what you are reading and be able to research yourself.

In addition, P1 had commenced building a STEM understanding:

> I think the most important thing with STEM is making sure it is interdisciplinary, and sort of a more holistic approach to thinking than we are learning.

On the other hand, P3 was beginning to connect her developing ideas about STEM with her experiences of the project and how she might approach her future teaching of STEM.

> I think that it's [mentor program] given me a much clearer understanding of, what is science? What are the STEM subjects? How I can integrate them better into the classroom? How I can motivate students? What technology I can use, what resources are available to the public? It's given me a lot more explicit direction to follow, rather than sort of the

> university education, which was much more focused on the curriculum; this is what the curriculum needs to teach. Then how do you translate that into a lesson? (P3)

One PSTs identified specific new knowledge she had gained:

> I have increased knowledge about coding, robots. (P1)

(ii) Self-perceptions as STEM educators

Initially, some of the PSTs expressed some uncertainty about their role in the program:

> STEM is a broad concept so there are so many avenues we might explore, even in the areas given, and we don't really know what level the kids will be working towards or where the lab techs will take us. (P2)

> I have a few ideas for what I want to run for my individual sessions during the student workshops. I've asked the other mentors what they want to run with no explicit response ... but I may not do it. (P1)

However, following the workshops, all of the mentors expressed more confidence in their roles as STEM educators and indicated some deeper thinking about their role in the translation of STEM for learners. For example:

> I was a lot more relaxed this time coming into it because I sort of knew what knew what to expect, knew what my basic roles were. (P3)

> I'm able to facilitate a bit more STEM thinking. (P2)

This building of PST confidence due to exposure, experience and opportunities to interact with STEM technicians, created a perceived increase in the STEM capacities of these PSTs, and may be interpreted as an indicator of STEM identity growth.

> Definitely a lot more confident. I think it does come from knowing what the programs are and having an idea and expectation of where I can interject with my knowledge or expertise. (P1)

Recognising the importance of good pedagogy in STEM learning also led several of the PSTs to comment on the teaching approach of the technicians. For example:

> The method of instruction was concerning – mostly this is what we're doing. This is how we do it. Now it's done ... Were the girls engaged? Not really. What would I do differently? Things like that. (P1)

One of the key issues to arise for this PST was the style of the workshops and the methods of instruction used by the technicians. The importance of schoolgirl engagement was mentioned and when asked about doing things differently themselves, a number of the PSTs recognised learner engagement as a problem of the workshops.

(iii) *Role as mentors*

An important purpose of including PSTs in this project was to provide both a 'near-peer' experience for the schoolgirls, with the aim of motivating the girls' interest, confidence and knowledge in STEM, as well as opening up PSTs' knowledge and confidence in STEM fields through working with each other (peer mentoring). However, it was clear that the PSTs often struggled to realise the full dimensions of the mentoring aspect of their role. Mostly, they tended to refer to themselves as a 'guide' or 'bridge' linking the school and university environments for the school teachers and students.

> I still feel like my first job is a guide, because I know where everything is. (P2)

> the people that are running the workshops at the moment are researchers ... or they're really scientifically minded. We bring the ability to teach students. So, I think that they can teach adults but they might not be able to teach 12-year-olds. So, I think that's what we bring, we bring the bridge. (P6)

By the end of the program, the PSTs had increased confidence in their roles as mentors, especially with regard to the pedagogical aspects of their role. For example,

> we're able to sort of ... interrupt, to say like, "Hey, just think about that thing that technician said ... what does that mean? (P2)

P1 explained that saw herself as both a bridge and a guide, a view that she maintained throughout the project:

> I still think that the primary goal [of the project] is to help with that bridging, especially for researchers like G, who is fantastic, but does not

> know how to bring it down to the level that is needed with high school students. So, I think that is a really important role that we have because it's something that the [school]teachers might not be comfortable to do that. But, I think the role of a mentor is also a bit of a guide. Somewhat of a role model but I do not know, it's hard. Yes, I think being passionate about science can help as well and having our background in science is really important too, because you know, there are a lot of teachers out there that do not. So, yes, I still think bridging the gap is the most important aspect of it all. (P1)

The participants generally also talked about their role as important in supporting the school girls' interest, motivation and self-confidence in STEM, opening up knowledge of opportunities in STEM fields, or improving perceptions of STEM subjects. Typical of this are comments such as:

> I do have an interest in fostering science in high school students – it is very important, even mathematics the way I see it. Probably more so mathematics since a lot of people are keen on science. It is opening up to new STEM areas. (P3)

And

> (I can) [h]elp identify misconceptions, especially if I do want to do things like break down the barriers of girls don't like math. I was like, "Well, that's not true." Not just from my previous experience, but also girls might have their own preconceptions and I can help with that. (P6)

By the end of the workshops, all of the PSTs reported that they felt confident in their role as mentors, mostly because they defined their role in terms of guiding or co-learning, rather than having full responsibilities for teaching the girls. They were also able to elaborate upon how their experiences and knowledge may be developed in terms of content or pedagogy in their own future teaching:

> I'm relatively confident that I'll be able to teach and give them help wherever possible. (P3)

The PSTs all indicated that the interaction with the schoolgirls made them feel "stronger and more confident" as STEM teachers and built their STEM identity both through mentoring and through role modelling:

> Definitely a lot more confident. I think it does come from knowing what the programs are and having an idea and expectation of where I can interject with my knowledge or expertise. That being said, all of this stuff is relatively new to me, what I am really comfortable with is nanotechnology because my background is in biology. But that is okay as well because you are being honest with the girls and saying, "Well, it's all new to me as well," is another one of those things that I think is important to show that science is all about everyone learning. (P2)

(*iv*) *Relationship between perceived purpose and participation*

An emerging theme through the interview data concerned the relationship between how the PSTs saw themselves in the project and the nature of their participation. Some PSTs used the experience as an opportunity to learn and develop new teaching skills, others saw it as a chance to bring STEM experiences to girls, others saw it as offering a potential advantage in future employment, while others saw it as an opportunity to become more involved with their peers (the latter very strong). It may have been that the voluntary nature of the mentoring contributed to less of a focus upon themselves as learners and greater focus upon the schoolgirls enjoying the workshops. Related to this, the PSTs may not have recognised the opportunities of the mentoring experience to build their own personal capacity and STEM identities. We elaborate below some examples of these purposes and perceived benefits in terms of university facilities, employability and working with others.

University Facilities

Three of the PSTs mentioned knowing about the university STEM facilities as a benefit to them.

> Knowing about RMIT STEM facilities and resources I have learnt (P1, 2, 3)

> just being able to liaise with the university more. (P2)

Given increasing emphasis on the role of STEM teachers as being able to locate and connect school-industry connections, these comments are noteworthy and may be worth emphasising in future iterations of the program.

Employability

Being able to distinguish herself as having something 'extra' to offer was identified by P3:

> Yes, it's been a big help just dropping the name of the program … it's been really good. So, getting interviews and getting a bit of a step forward. (P3)

We also noted, as facilitators, that developing an extended relationship with a school meant that PSTs had greater opportunities to 'showcase' themselves to potential employers. In fact, one of the mentor PSTs was employed in the school she was working with, following the program.

Working with Others

PSTs frequently mentioned that they greatly appreciated the opportunity to work with their peers, the participating school teachers and the STEM university technicians throughout the research project. However, they did find that working with the technicians was challenging at times because of the highly complex ways in which they talked about their work, that made it difficult to connect with their own experiences and ideas, and also (even more so) with the schoolgirls.

> M [technican] is great, but the level is just too high. Once you become so engrossed in your own research, the terminology used becomes really normal to you, so you kind of forget that people do not necessarily know what you are talking about … complicated things seem easy to them. (P1)

The role of the school based teachers seemed to have an important influence on PSTs capacity to step into their STEM educator role. For example, P1 commented on her experiences of working with teachers from different schools:

> it's good to see that the teachers at [school] … have taken a step back …, because it helps us as mentors to further develop …. I found last time working with [name] teacher, she was just difficult … Did not allow for much mentoring at all. So, being able to happily interject and interact with the students has been much easier this time because the teachers themselves are I think a bit more open to the program. (P1)

Through their reported experiences of the program, it was clear that initially, the PSTs identified their role only in terms of the university workshops with the schoolgirls. However, by the end of the program, they were more able to appreciate working with each other, and with the STEM technicians and teachers. As Archer et al. (2010) noted, the opportunity to work with peers and experts is significant in the development of a STEM identity. The data arising from this research indicates this development and supports the role of peers and technicians in promoting this STEM identity development.

Both Sanders (2009) and Schmidt and Fulton (2016) have identified confidence in STEM as critical to building more effective educators in STEM. The data from this project indicated that how the learners/schoolgirls and their schools defined mentoring was also a variant. The commitment of the schools and the schoolgirls involved in the project also impacted the confidence of the PSTs, however we did not collect data about this aspect. The confidence and STEM self-perception of the PSTs was also impacted by the organisation of the mentoring situation. On a number of occasions mentors found it difficult to find time to meet with the STEM technicians and whilst the STEM technicians provided feedback to the organisers that they felt very satisfied with their interactions with the schoolgirls, the role of the mentor was interpreted differently by different technicians. For example, some held on to their status as experts and did not enable the PSTs to share in the activities of the workshops beyond 'crowd control.' We consider that this impacted upon the pre-service teacher's self-confidence and self-perceptions. Sjaastad (2012) identifies *models* as parents, teachers and others displaying a STEM professional identity – and *definers* – those helping a young person in the process of setting goals, defining values and identifying personal strengths in their educational choices. In this study, the PSTs were both models and definers, serving as models for the schoolgirls through enacting their STEM identity, and also being the definer through being acutely aware of the significance of defining values and setting goals with the schoolgirls. In one way, by interchanging these two roles, the PSTs also extended their confidence through interaction with the schoolgirls and thus extended their own STEM identities and thus their own STEM identities.

8 Conclusion

Glavich (2016) notes there is an obvious need for strong STEM preservice teacher programs. Without attending to the root of the problem, beginning teachers will not be able to branch out and implement STEM-focused curricula. Universities are central to improving PSTs' understanding of, and interest in STEM disciplines.

> Australia is looking to universities to re-imagine what STEM education should be, and re-engineer for a future in which this is central. (Prinsley et al., 2015, p. 1)

As far as what can be learned from this project, there are obvious limitations in terms of the small sample size and limited data sources. Thus, the claims we make must be modest and in some cases, speculative. Nevertheless, we believe that important implications can be drawn from both the design of the program and the insights that can usefully inform STEM education, as we outline below.

We contend that teacher education institutions can take a lead in providing opportunities for developing PST's perceptions and understandings of STEM, as once embedded in schools and classrooms, teacher identities tend to become more fixed and stable. Outcomes of this research project focused upon the main themes of: mentors' understandings of STEM and developing a pedagogy around STEM, their self-identity and learning in STEM, and the value of the research project for teachers in preparation. Through this mentoring project small changes in STEM conceptualisation and STEM self-identities occurred in the PSTs involved and evidence of increased confidence in their capacities as STEM educators also emerged. This building of confidence due to exposure, experience and relating to STEM technicians, created a perceived increased level of STEM capacity in the PSTs. This building of their confidence and perception of themselves as STEM educators growing is an indicator of STEM identity growth. The limited data and time scale means that we can only 'postulate' on this identity aspect, based on these indicators.

Although the research project only involved a small number of PSTs and such numbers create a limitation upon broader generalisations, there is evidence that mentoring projects such as this can create opportunities to improve the STEM capabilities and perceptions of pre-service teachers. The research project is thus of significance to those involved in the preparation of teachers and the wider STEM community, as society grapples with the task of preparing young people, especially young women, for a STEM-inspired future.

References

Archer, L., DeWitt, J., Osborne, J., Dillon, J., Willis, B., & Wong, B. (2010). 'Doing' science versus 'being' a scientist: Examining 10/11-year-old schoolchildren's constructions of science through the lens of identity. *Science Education, 94*(4), 617–639. doi:10.1002/sce.20399

Barab, S. A., & Hay, K. E. (2001). Doing science at the elbows of experts: Issues related to the science apprenticeship camp. *Journal of Research in Science Teaching, 38*(1), 70–102.

Breiner, J. M., Johnson, C. C., Sheats Harkness, S., & Koehler, C. M. (2012). What is STEM? A discussion about conceptions of STEM in education and partnerships. *School Science and Mathematics, 112*(1), 3–11. doi:10.1111/j.1949-8594.2011.00109.x

Carlone, H. B., & Johnson, A. (2007). Understanding the science experiences of successful women of color: Science identity as an analytic lens. *Journal of Research in Science Teaching, 44*(8), 1187–1218. doi:10.1002/tea.20237

Casey, K. M. A., & Shore, B. M. (2000). Mentors' contributions to gifted adolescents' affective, social, and vocational development. *Roeper Review, 22*(4), 227–230.

Chiu, A., Price, C. A., Ovrahim, E. (2015, April 11–14). *Supporting elementary and middle school STEM education at the whole school level: A review of the literature.* Paper presented at NARST 2015 Annual Conference, Chicago, IL.

Glavich, C. (2016). Growing strong STEMs: Reflections of a beginning teacher's preservice program. *Issues in Teacher Education, 25*(2), 89.

Hobbs, L., Cripps Clark, J., & Plant, B. (2017). Successful students STEM program: Teacher learning through a multifacted vision for STEM education. In R. Jorgensen & K. Larkin (Eds.), *STEM education in the junior secondary* (pp. 133–168). Singapore: Springer.

Jaipal-Jamani, K., & Angeli, C. (2017). Effect of robotics on elementary preservice teachers' self-efficacy, science learning, and computational thinking. *Journal of Science Education and Technology, 26*(2), 175–192.

Johnson, C. C. (2013). Conceptualizing integrated STEM education. *School Science and Mathematics, 113*(8), 367–368.

Kim, C., Kim, D., Yuan, J., Hill, R. B., Doshi, P., & Thai, C. N. (2015). Robotics to promote elementary education pre-service teachers' STEM engagement, learning, and teaching. *Computers & Education, 91*, 14–31. doi:10.1016/j.compedu.2015.08.005

Lave, J., & Wenger, E. (1991). *Situated learning: Legitimate peripheral participation.* Cambridge: Cambridge University Press.

Le Cornu, R. (2005). Peer mentoring: Engaging pre-service teachers in mentoring one another. *Mentoring & Tutoring: Partnership in Learning, 13*(3), 355–366. doi:10.1080/13611260500105592

McLaughlin, P., & Berry, A. (2017). *STEM in situ: Imagining entrepreneurial futures.* Australia: National Innovation and Science Agenda (NISA).

Nadelson, L. S., Callahan, J., Pyke, P., Hay, A., Dance, M., & Pfiester, J. (2013). Teacher STEM perception and preparation: Inquiry-based STEM professional development for elementary teachers. *The Journal of Educational Research, 106*(2), 157–168. doi: 10.1080/00220671.2012.667014

Office of the Chief Scientist. (2016). *Australia's STEM workforce: Science, technology, engineering and mathematics.* Canberra: Australian Government.

Polman, J. L., & Miller, D. (2010). Changing stories: Trajectories of identification among African American youth in a science outreach apprenticeship. *American Education Research Journal, 47*, 879–918. doi:10.3102/0002831210367513

Prinsley, R., & Baranyai, K. (2015). *STEM-trained and job-ready*. Canberra: Office of the Chief Scientist.

Radloff, J., & Guzey, S. (2017). Investigating changes in preservice teachers' conceptions of STEM education following video analysis and reflection. *School Science and Mathematics, 117*(3–4), 158–167. doi:10.1111/ssm.12218

Sadler, D. R. (2010). Beyond feedback: Developing student capability in complex appraisal. *Assessment & Evaluation in Higher Education, 35*(5), 535–550. doi:10.1080/02602930903541015

Sanders, M. (2009). STEM, STEM education, STEMmania. *The Technology Teacher, 68*(4), 20–26.

Schmidt, M., & Fulton, L. (2016). Transforming a traditional inquiry-based science unit into a STEM unit for elementary pre-service teachers: A view from the trenches. *Journal of Science Education and Technology, 25*(2), 302–315.

Sjaastad, J. (2012). Sources of inspiration: The role of significant persons in young people's choice of science in higher education. *International Journal of Science Education, 34*(10), 1615–1636. doi:10.1080/09500693.2011.590543

Stout, J. G., Dasgupta, N., Hunsinger, M., & McManus, M. A. (2011). STEMing the tide: Using ingroup experts to inoculate women's self-concept in Science, Technology, Engineering, and Mathematics (STEM). *Journal of Personality and Social Psychology, 100*(2), 255. doi:10.1037/a0021385

Tenenbaum, L. S., Anderson, M. K., Jett, M., & Yourick, D. L. (2014). An innovative near-peer mentoring model for undergraduate and secondary students: STEM focus. *Innovative Higher Education, 39*(5), 375–385. doi:10.1007/s10755-014-9286-3

UNESCO. (2017). *Cracking the code: Girls' and women's education in Science, Technology, Engineering and Mathematics* (*STEM*). Retrieved February 5, 2018, from http://unesdoc.unesco.org/images/0025/002534/253479E.pdf

Vasquez, J. A. (2014). STEM—Beyond the acronym. *Educational Leadership, 72*(4), 10–15.

Wai-Ling Packard, B., & Nguyen, D. (2003). Science career-related possible selves of adolescent girls: A longitudinal study. *Journal of Career Development, 29*(4), 251–263.

Wang, M. T., & Degol, J. (2013). Motivational pathways to STEM career choices: Using expectancy-value perspective to understand individual and gender differences in STEM fields. *Developmental Review, 33*, 304–340. doi:10.1016/j.dr.2013.08.001

Windschitl, M. (2004). Folk theories of "inquiry:" How preservice teachers reproduce the discourse and practices of an atheoretical scientific method. *Journal of Research in Science Teaching, 41*, 481–512. doi:10.1002/tea.20010

CHAPTER 12

Building Academic Leadership in STEM Education

Tricia McLaughlin and Belinda Kennedy

Abstract

Whilst much of the emphasis upon STEM in Australia has focussed upon the need for greater learning opportunities for students and emerging skill gaps, little attention has been directed towards the academic workforce and their capacity to deliver STEM education in tertiary contexts. This chapter reports on a nationally funded Australian Government project in building capacity for academics from STEM and other disciplines to engage in cross-disciplinary activities. Two of the national case studies are selected for discussion in this chapter. In these case studies, staff awareness and confidence in STEM cross-disciplinary work increased, and their understanding of the value of such cross-disciplinary work for students also increased. These case studies provide one model of ensuring that academic leadership is at the forefront of STEM learning in the future.

Keywords

STEM – academic leadership – ecosystem

1 Introduction

> The global economy is changing. New technologies and smart companies lead. New industries and new sources of wealth are emerging. New skills are required for work at all levels. Australians must decide whether we will be in the forefront of these changes or be left behind. (Chief Scientist, 2016).

With identified current and future skills shortages across the STEM professions in Australia (OCS, 2016), the Australian Government's Office of Learning and Teaching through its Leadership for Excellence in Learning and Teaching Program, supported an investigation into systematic, structured and sustainable models of STEM academic leadership in higher education.

 | DOI:10.1163/9789004391413_013

The STEM Ecosystem project, as it was called, was led by researchers at RMIT University in a consortium of three other universities (The University Sydney, University Southern Queensland, and The University Queensland). In framing the investigative project, the consortium had identified the need for academic staff from science, technology, engineering and mathematics (STEM) disciplines to impart knowledge from their own disciplines across new STEM learning boundaries and build skills in providing cross disciplinary STEM opportunities for students. Through the connection of industry, STEM academics and students, the project attempted to create these cross-disciplinary STEM opportunities and demonstrate the emerging potential in cross-disciplinary learning and teaching in higher education.

The primary objective of the STEM Ecosystem project was to foster leadership potential for mathematics, technology, science and engineering staff and to build staff capacity and confidence in STEM cross-disciplinary pedagogy. The specific objectives were to:

- Increase capacity of STEM academic staff to design, develop and lead industry-relevant cross-disciplinary courses
- Engage STEM discipline learning and teaching staff with the Ecosystem
- Increase the number of STEM staff-initiated cross-disciplinary learning and teaching projects in each institution
- Improve understanding and awareness of specific learning and teaching strategies to maximise the outcomes for students engaging in STEM cross-disciplinary projects
- Embed cross-disciplinary teaching and learning strategies in discipline curriculum

The STEM Ecosystem project was built upon the concept of skills ecosystems and adopted the methodological approach refined in the work of Buchanan (2006). The use of Ecosystems as methodological approaches for the mentoring of leadership projects and the promotion of skill networks is well grounded in the literature. Skills Ecosystems provide leadership opportunities for those involved in them. Introduced by Finegold (1999) in his examination of knowledge and skill creation and transmission in the cluster of computer and biomedical firms in the Silicon Valley, USA, the methodological approach has been used in and across various industries as a self-sustaining management model (Buchanan, 2006; Payne, 2008; Anderson & Warhurst, 2012).

The Ecosystem concept has also been applied to environments that foster entrepreneurial skill development and those that foster regional, national and international networks of potential leaders (Hall & Lansbury, 2006; Schwalje, 2011; Buchanan, Anderson, & Power, 2017). Since early 2000, the concept of skills Ecosystems as leadership catalysts has been recognised by Australian Governments, with funding in 2003 to a Skills Ecosystem national program of joint state-federal developments that tested new and dynamic frameworks for leadership in selected industries. One outcome of the national program was partnership projects between tertiary institutions and industry that focussed upon improving staff capacity. A key finding of the nationally funded projects was their ability to highlight aspects of the Ecosystem environment that influenced the development, application and replenishment of skills and leadership opportunities for staff working within the Ecosystem (DETA, 2007; Buchanan et al., 2017).

The role of skills ecosystems and their success in sustained change by exposing individuals to new forms of work and leadership roles within a supportive workplace environment has also been identified in specific Australian contexts by Buchanan, Baldwin, and Wright (2011). Building upon this literature and the previous learning and teaching innovations of the project team members, this project used the skills Ecosystem model to bring together STEM discipline academics to generate positive, mutually reinforcing dynamics to fuel ongoing knowledge, creation and growth of personnel into leaders able to provide learning and teaching direction for STEM cross-disciplinary projects.

There is evidence that best practice approaches to STEM learning and teaching ensure that students not only acquire knowledge, but also learn how to apply and adapt this knowledge in a variety of contexts (OCS, 2016; O'Hara et al., 2017). However, learning and teaching in the science, technology, mathematics and engineering disciplines often remains narrowly focused and content entrenched, especially in tertiary education. Rice (2011) notes that STEM disciplines in tertiary education are often taught through paradigms which are not practical, nor reflective of real-life industry. STEM disciplines are seen as opportunities to induct students into the content of the discipline, not as opportunities to develop cross-disciplinary skills or develop solutions to complex future problems. Efforts by STEM academics to undertake cross-disciplinary industry-based projects are rare and often not sustained. Silos of best practice cross-disciplinary projects remain at the fringes of the curriculum, often in the "project, competition or challenge" arena, and are not capitalised upon at the institutional level or at the national level for the benefit of other tertiary STEM staff.

The Office of Chief Scientist (2016) has identified a range of factors contributing to academic staff not engaging in cross-disciplinary STEM in tertiary education:
- Specific discipline skill shortages,
- Lack of confidence,

- Rapid technological change,
- Lack of role-modelling by experienced academics, and
- Discipline-specific learning and teaching demands of tertiary structures.

The cross-disciplinary STEM work that is undertaken often happens in isolation, with STEM cross-disciplinary learning and teaching ad-hoc and not formally embedded at the institutional level (Rice, 2011).

Increasingly the needs of employers and future work opportunities do not recognise boundaries of discipline-specific education. Preparing students for new ways of dealing with growing bodies of knowledge that no longer fit neatly into a discipline program creates enormous challenges for tertiary institutions organised along strict STEM discipline lines. Global industries require individuals with skills and knowledge across a range of STEM disciplines. The evidence shows that high performing countries in terms of STEM advances have a reliable pipeline of STEM graduates, whose cross-disciplinary skills are valued by employers (OECD, 2011). Restricting learning in tertiary institutions to the specifics of one discipline is limiting options for Australian graduates.

The commonality of threshold learning outcomes across STEM disciplines alongside growing industry requirements for such skills and knowledge also creates a compelling need for cross-disciplinary STEM learning in higher education (OCS, 2016). However, trialling and embedding cross-disciplinary approaches in STEM learning into core undergraduate discipline curricula is complex for most universities. Traditional organisational structures of tertiary institutions mean that students (and academic staff) have little opportunity for cross-disciplinary networking or learning. Whilst there is some emerging research that documents the positive effects of cross-disciplinary approaches among STEM disciplines upon the students' achievement, satisfaction and employability (Pang & Good, 2000), the understanding of the need and emphasis lacks traction in most Australian universities. More recent research has focussed upon the pedagogy behind cross-disciplinary learning. Kavanagh (2011) notes that such courses capture students' intellectual interest, prepare students for work by developing higher-order cognitive skills, and increase students' tolerance for ambiguity, sensitivity to ethical issues, and creativity. Lotz-Sisitka, Wals, Kronlid, and McGarry (2015) in discussing the role of tertiary education across educational boundaries, define pedagogy as being a series of roles, where learning is that of 'interpretive actors' who fight, negotiate, search, and learn with others.

But although research on the integrative learning and teaching approaches amongst STEM disciplines has grown, there are still a number of serious practical challenges to engaging cross-disciplinary best practice. STEM academics and their implementation of learning and teaching cross-disciplinary projects depends greatly on their perceptions of the advantages of such an approach, the organisational context and the leadership and support received in their teaching environment. That is, the decision by STEM academics to implement cross-disciplinary learning and teaching approaches to improve student outcomes engagement and future STEM opportunities is heavily dependent upon leadership support and role-modelling of such practice (McLaughlin & Kennedy, 2016).

By adapting the Ecosystem leadership concept to STEM academic staff, the project would be able to build upon existing research and provide learning and teaching leadership for grassroots STEM academics, the wider institutions and the tertiary sector. The project would utilise the key features of skills ecosystems to engage STEM discipline staff as cross-disciplinary leaders and create an ecosystem of interdependent networks and collaborative leadership frameworks under the STEM banner.

2 The Skill Ecosystem Case Studies

The four key features of an Ecosystem that are necessary for successful innovation, development change and dissemination (Finegold, 1999; Buchanan et al., 2017) formed the methodological stages of this project:

- A supportive host environment, where ideas and early career teaching and learning staff can be mentored and test new ideas;
- A catalyst for "being" and "remaining" through the cross-disciplinary courses;
- Continual nourishment or cross-fertilisation of ideas to encourage and stimulate change and development; and
- A high degree of independence, free from previous or existing discipline silos.

The STEM Ecosystem case studies were situated in diverse disciplines at the project universities. Data for the case studies was collected from teaching staff, industry mentors and students enrolled in the cross-disciplinary work to understand the learning and teaching outputs of the cross-disciplinary activity. This data was collected via semi-structured interviews and in survey format. Academic staff and students who agreed to be involved in this research answered a series of questions about the learning activities and their opinion of the value of such learning activities. Analysis was undertaken both quantitatively and qualitatively and by an independent project manager, using software and survey tools. The survey relied on aggregate data only and the semi-structured interviews on individual student and staff responses. Two of the key case studies are commented upon in this chapter: "the WIC and the Ecocities Symposium."

3 A Catalyst and Supportive Host Environment

The first phase of the Ecosystem, was the creation of the STEM cross-disciplinary workshops or courses by academic staff and industry partners. The Water Innovation Challenge (WIC) at RMIT was the case study selected to illustrate the creation of a catalyst and host environment for STEM academic staff and students. The WIC created opportunities for staff and students from four different discipline-based RMIT schools to work alongside industry practitioners in a cross-disciplinary-skilled team to design, build and present innovative water sanitation solutions for a selected Bangladesh community. The project client responsible for the community selection and installation was Health Habitat, a global Non-Government Organisation operating across many third world countries. The WIC aims were:

- To realise a viable water sanitation solution for a Bangladesh community
- To meet client needs in tender documentation (CAD, Budgeting, solutions etc), and
- To showcase solutions at a practical 3 day challenge.

The WIC was conducted intensively over an eight week period in semester 1, 2014. Students were invited from science, engineering and health schools. Staff involved in the WIC represented engineering, health, plumbing, media and IT disciplines. Staff and students were given the challenge, resources, information and worked in one large team with needs-based sub-groups formed and reformed as the project progressed. Academic staff contributed knowledge, and disciplinary learning. Staff in this project were surveyed and a smaller number (4) participated in semi-structured interviews asking them to elaborate on the survey questions. The collection of this data took place post-project. The interview responses and the answers to the written survey were recorded. In the three survey questions, participants were asked to respond simply with "yes," "no' or "don't know." Table 12.1 shows a summary of the staff survey responses. In other published papers, student data has also been reported.

Academic staff involved in this project overwhelmingly agreed that the cross-disciplinary projects were worthwhile and prepared students for work upon graduation. Comments included:

TABLE 12.1 Staff responses to survey questions (response rate: 100%; n=14)

Survey question	Yes	No	Unsure
Did you like working with students from different disciplines and levels of study to your normal cohort?	95%	0%	5%
Did you think that this project has prepared students for work upon graduation?	95%	0%	5%
Would you undertake similar types of cross-disciplinary projects in the future as an academic?	100%	0%	0%

> By allowing students to see how the various pieces of a project come together and the bigger results, they can grasp the value of what their parts are required to do.

And,

> Interestingly I learnt more from the students than they did from me – the engineering students were just great – they gave me real insights into what we were trying to design and why.

The WIC nurtured and created confidence across staff from a range of disciplines, including early career academics. This impact upon staff was intentional – a key element of the OLT-funded STEM Ecosystem was the nurturing of academics through developing opportunities for them to lead in cross-disciplinary STEM arenas. This was achieved as the comments from staff involved in the project indicate:

> Without being exposed to these new experiences or being out of my comfort zone I don't think that it's possible for me to develop personally or professionally.

Through this case study, the Ecosystem was established and built upon the experiences, knowledge and leadership of the STEM academics to maintain interest and a small community of practice. Staff, including an early career academic used the WIC to test new ideas and refine learning and teaching experiences to enable dissemination and modelling for other STEM academics in other universities. The STEM academics were able to draw from the Ecosystem through mentoring and peer support.

4 Support and Independence – Diaspora

The second phase of the Ecosystem was to support the STEM academic leaders as they further developed and embedded information, advice and feedback about the cross-disciplinary projects. This phase focussed upon the formal recognition of this leadership through academic publications and conference presentations. Throughout this phase further development of the STEM academics as leaders was continued through mentoring and their involvement in a community of practice. This nurturing and sustainability is evident in the EcoCities case study.

"EcoCities" formed part of the STEM Ecosystem project and presented a new model of understanding and applying STEM knowledge for students from a range of disciplines. The "EcoCities" symposium was conducted over a three day period at RMIT in large, interactive learning spaces with over 150 industry, undergraduate and postgraduate students and academics working on "live" industry problems in urban ecology. The concept behind the symposium was to promote a stimulating, interactive learning environment for students from a range of disciplines to gain from exposure to STEM industry experts and academics in a "real-world" work environment. Industry facilitators from urban ecology disciplines as diverse as transport and food conducted workshops where students and industry attendees participated in defining problems within the over-riding constraints and boundaries of the City of Melbourne. Workshop groups worked through specific urban ecology problems such as waste management in a large city, food storage and supply, sustainable living, transport and logistics to arrive at real-world, credible and sustainable solutions. Workshops were of varying duration depending upon the problem, the solution and the capacity of the participants. Solutions were allowed time and space to be presented.

The aims of the Ecocities Symposium were:

- To establish a cross-disciplinary learning future in Urban Ecology at RMIT University, with international connections
- To bring together Melbourne city leaders with internationally renowned teaching and research experts at an international symposium.

The selection of national and international thought leaders in urban ecology to speak and share their ideas at the symposium was critical to the success of this Ecosystem model. The international experts were the principal facilitators forming the nucleus of the ecosystem supplemented by interested staff, students and local industry experts. The international experts and academic staff created master classes in the symposium. Students were encouraged to select master classes, demonstrations or workshops based around their interests and skill contributions. While the workshops/masterclasses were of differing intensity and duration, all involved industry experts, academics and students discussing issues on the various particular topics. Solutions to the various workshop themes were tackled in the design studios. Staff and students from engineering, applied science, architecture, planning, health, biodiversity, social science and management, economics and business, culture and heritage, arts and property services and the wider university community attended and contributed to the Symposium.

Using the Ecosystem model of support and independence led to diverse and informed discussions. Staff were surveyed at the end of the symposium and their responses to the same set of questions as the WIC were recorded. All questions required a written response and the results are recorded in Table 12.2.

Staff involved in this case study enjoyed being leaders of cross-disciplinary workshops and 100% intended to offer such workshops/master classes again. A smaller number of staff (4) participated in semi-structured interviews asking them to elaborate on the survey questions.

On the issue of working with different disciplines, staff noted:

> Leadership from industry experts gave students the confidence to tackle unfamiliar problems.

And,

> It was a great opportunity to further develop my ideas and hear from experts across other disciplines as to the value of such collaborations.

The Ecocities symposium will be the catalyst for many outcomes and future spin-offs for the lead University. Evidence from the staff data and the industry

TABLE 12.2 Staff responses to survey questions (response rate: 100%; n=8)

Survey question	Yes	No	Unsure
Did you like working with students from different disciplines and levels of study to your normal cohort?	85%	0%	15%
Did you think that this project has prepared students for work upon graduation?	90%	0%	10%
Would you undertake similar types of cross-disciplinary projects in the future as an academic?	100%	0%	0%

and student data elsewhere analysed indicated that the symposium achieved the following outcomes:

- Launched a new cross-disciplinary learning activity/focus for the University
- Provided content to seed further coursework components across the university programs
- Enhanced collaboration with industry in the learning and teaching environment

The key features of an Ecosystem, necessary for successful innovation, development of change and dissemination (Finegold, 1999), were present in the Urban Ecology "Ecocities" Symposium:

1. The continual nourishment or cross-fertilisation of ideas

The three day nature of the symposium allowed for continual nourishment of ideas and solutions. The coming together of differing disciplines and backgrounds meant new perspectives that encouraged and stimulated possibilities for change within the workshops and the eventual solutions. Finally, the input of the industry "experts" produced constant cross-fertilisation of ideas and knowledge.

2. A high degree of independence, free from previous or existing discipline silos

By releasing students and academic staff from the constraint of set discipline learning criteria, the symposium allowed application of knowledge and previous learnt information. It introduced problem solving skills in a tight solutions-focussed framework and relied upon new applications of discipline knowledge. Students were compelled to communicate and participate in professional debate around their discipline and across the boundaries of other areas. Reflective practice, critical analysis and trial and error skills were critical to success.

5 Conclusion

A large number of case studies across the four universities were trialled as part of the STEM Ecosystem. Two of these case studies have been reported in this chapter. Surveys of staff in these case studies have illustrated the relevance and importance of the following;

- The provision of opportunities for current STEM staff to demonstrate, lead and collaborate by utilising existing STEM courses in cross-disciplinary projects showing the commonality of STEM discipline skills
- The enabling of confidence and capacity-building in STEM academics, including early career academics

- The creation of national leadership recognition of cross-disciplinary STEM learning and teaching

This project, by providing an Ecosystem of STEM academics who were able to lead, encourage, demonstrate and mentor staff in best practice cross-disciplinary learning and teaching, was an inspiration for innovation and change within STEM-related disciplines. Academic staff were supported and encouraged to move away from "paradigmatic examples" of STEM teaching (Rice, 2011) and through the confidence gained by peer and senior academic support, trial new approaches in their own institutions. Evidence from the two illustrated case studies here shows staff were empowered by their involvement in the Ecosystem and the industry mentoring opportunities. The Ecosystem model was a mesh network of inter-connected individuals, who drew skills and knowledge from role-modelling and mentoring, and who trialled new approaches to effect change in their own teaching, and also inspire and motivate others to change.

Australia's STEM tertiary academics need to be equipped to deliver course content with confidence and inspiration and develop all students to their full potential. Curricula and assessment criteria should prioritise curiosity-driven and problem-based learning of STEM—STEM as it is practised—alongside the cross-disciplinary knowledge that STEM requires. The tertiary education system must ensure that students not only acquire specific discipline knowledge, but also learn how to apply and adapt this knowledge to a variety of contexts.

However, the successful adoption of the Ecosystem and the showcase case studies by Australian universities involved a number of challenges. These included:

- Level of enthusiasm, confidence and experience of academic staff not directly involved in the project to engage in the process
- Level of understanding of the importance of pedagogy in STEM cross-disciplinary projects
- Appropriate administrative support and resources being available to assist with the daily running of the projects
- The willingness of academic staff to balance their workload to implement the cross-disciplinary courses/workshops

STEM skills are critical to the economic management and success of Australia. STEM skills are the lifeblood of emerging knowledge-based industries such as biotechnology, information and communications technology (ICT) and advanced manufacturing and provide competitive advantage to established industries such as agriculture, resources and healthcare. These are all industries and global networks to which Australian graduates will be drawn.

An education in STEM fosters a range of generic and quantitative skills and ways of thinking that enable individuals to see and grasp opportunities. These capabilities including deep knowledge of a subject, creativity, problem solving, critical thinking and communication skills are relevant to an increasingly wide range of occupations. They will be part of the foundation of adaptive and nimble workplaces of the future. The STEM Ecosystem project demonstrated leadership in STEM education by using these capabilities as the basis for cross-disciplinary teamwork projects that have enhanced student discipline learning and built capacity in STEM discipline academics.

STEM is an important part of Australia's development. This project expanded the reach of STEM in universities and across tertiary staff in this country. But a thorough STEM strategy is about far more than simply igniting a passion for STEM in academia. It is about applying STEM skills to enhance capacity, which will translate into direct benefits for society through improvements in economy, new opportunities for industries and advances in the standard of living. This project illustrated that changes in STEM demand commitment and collaborative effort across staff in tertiary education to succeed.

References

Anderson, P., & Warhurst, C. (2012). *Lost in translation? Skills policy and the shift to skill ecosystems*. In T. Dolphin & D. Nash (Eds.), *Complex new world: New era economics* (pp. 109–120). London: Institute for Public Policy Research (IPPR).

Buchanan, J. (2006). *From 'skill shortages' to decent work: The role of skill ecosystems*. Sydney: NSW, DET.

Buchanan, J., Anderson, P., & Power, G. (2017). Skill ecosystems. In *The Oxford handbook of skills and training* (pp. 444–465). Oxford: Oxford University Press.

Buchanan, J., Baldwin, S., & Wright, S. (2011). *Understanding and improving labour mobility: A scoping paper*. Paper presented at the Workplace Research Centre, The University of Sydney Business School, NCVER, Adelaide, Australia.

DETA. (2007). *Towards a 10 year plan for Science, Technology, Engineering and Mathematics (STEM) education and skills in Queensland—Discussion paper*. Retrieved January, 2011, from http://education.qld.gov.au/projects/stemplan/docs/stem-discussion-paper.pdf

Finegold, D. (1999). Creating self-sustaining high-skill ecosystems. *Oxford Review of Economic Policy, 15*(1), 60–81.

Hall, R., & Lansbury, R. (2006). Skills in Australia: Towards workforce development and sustainable skill ecosystems. *Journal of Industrial Relations, 48*(5). Retrieved from https://doi.org/10.1177%2F0022185606070106

Kavanagh, L., & Cokley, J. (2011). A learning collaboration between engineering and journalism undergraduate students prompts interdisciplinary behavior. *Advances in Engineering Education, 2*(3), 1–22.

McLaughlin, P., & Kennedy, B. (2016). *The STEM ecosystem-building cross disciplinary leadership capacity in Science, Technology, Engineering and Mathematics.* Melbourne: Chapman Publishing.

Office of the Chief Scientist. (2016). *Australia's STEM workforce: Science, Technology, Engineering and Mathematics.* Canberra: Australian Government.

O'Hara, S., Pritchard, R., Pitta, D., Newton, R. N., Do, U. H., & Sullivan, L. (2017). Academic Language and Literacy in Every Setting (ALLIES+): Strengthening the STEM learning ecosystem. In *Science teacher preparation in content-based second language acquisition* (pp. 199–213). Cham: Springer. Retrieved from https://doi.org/10.1007/978-3-319-43516-9_11

Organisation for Economic Co-operation and Development (OECD). (2011). *Over-qualified or under-skilled: A review of existing literature.* Paris: OECD.

Pang, J. S., & Good, R. (2000). A review of the integration of science and mathematics: Implications for further research. *School Science and Mathematics, 100*(2), 73–82.

Payne, J. (2008). Skills in context: What can the UK learn from Australia's skill ecosystem projects? *Policy & Politics, 36*(3), 307–323.

Reiss, M., & Holman, J. (2007). Blurring the boundaries: STEM education and education for sustainable development. *Design and Technology Education, 14*(1), 37–45.

Rice, J. (2011). *Good practice report: Assessment of Science, Technology, Engineering and Mathematics (STEM) students.* Sydney: ALTC Ltd.

Schwalje, W. (2011). *A conceptual model of national skills formation for knowledge-based economic development* (MPRA Paper No. 30302). Munich: University Library of Munich.

CHAPTER 13

Epilogue: What Now for STEM?

Linda Hobbs

1 Introduction

In 2018, Science, Technology, Engineering and Mathematics (STEM) education is in its formative years in countries like Australia (Timms, Moyle, Weldon, & Mitchell, 2018), as has been demonstrated in this book. There appears to be enough of a groundswell of interest in all education sectors for STEM to be moving beyond the rhetoric of what needs to be done to actual practice. While there is no STEM curriculum per se in Australia, and teachers are still working out what STEM can mean for them in their schools, there are now many examples of STEM innovation that teachers can seek out to guide local decision-making (see, for example, Australian Industry Group, 2016). The task now is to document, work together, and learn from each other.

This book makes a start on this. The key learnings offered in this book cover a range of issues: the broad discussion of what STEM is and can be (conceptualizations of STEM); preparing pre-service teachers (PSTs) as teachers of STEM (STEM in initial teacher education); how schools can introduce powerful STEM learning (STEM in schools); and making tertiary learning more focused on STEM (STEM at tertiary level). In the below I do not attempt to summarise the chapters in their entirety, as such summaries are provided in the Introduction, but I intend to highlight the key learnings offered in relation to each of these issues as they arise from the work represented here. I then propose a set of principles for effective STEM education, which include important considerations when bringing STEM to life in our schools and universities.

2 Conceptualizations of STEM

As discussed in the Introduction more generally, STEM is undergoing an identity crisis. Chapters 1 and 3 provide insightful discussions of the meanings attached to STEM in Australia. Along with a number of other authors in this book, in Chapter 1, Fraser, Earle and Fitzallen have focused on the cross-disciplinary opportunities that STEM affords (disciplinary as distinct from multi-, inter- and

 | DOI:10.1163/9789004391413_014

trans-disciplinary approaches). In Chapter 11 Berry, McLaughlin and Cooper provided a useful framework from Radloff and Guzey (2016) that shows how this emphasis on an integration and interdisciplinarity continuum is only one of three ways that STEM education is being promoted; the second being as an instructional focus where generic competencies are used to make connections between schools, community, and industry; and the third being STEM focused on stakeholders' needs and contexts, like solving industry-based problems. These diversity of approaches, ways of conceptualizing and involvement of different subjects and groups of people, give schools a variety of ways forward when deciding how to incorporate STEM, where to place the emphasis on learning, and who to get involved.

Fraser et al., however, point out that there is a need for data to show the effects of each model on student learning, aspirations, and career choices. They also echo the concerns raised by a number of other commentators of the need for deep understanding of the STEM disciplines being integrated (Honey, Pearson, & Schweingruber, 2014), in particular to "understand how they are connected and to make those connections explicit for students". They suggest that this can be achieved by teachers working within inter-disciplinary teams; and that if we expect students to achieve some level of understanding of the disciplines being integrated, then we should expect teachers to achieve a similar degree of understanding.

In Chapter 3, Jordan provides commentary on STEM as a national agenda and an educational priority. According to this analysis STEM is represented in policy documents in Australia as: (1) national enterprise, (2) sustaining economic growth, (3) maintaining prosperity, (4) not being left behind, (5) securing a workforce, and (6) declining. According to this analysis, increasing participation and performance in STEM is represented to be relatively straightforward. However, the analysis is both insightful and helpful in highlighting the silences in these documents, that is, the relative complexities involved in the STEM enterprise.

3 STEM in Initial Teacher Education

The initiatives being reported in Chapters 9, 10 and 11 all respond to the need to prepare teachers for the reality of having to attend to STEM when teaching in schools. An analysis of PST perceptions of STEM by Cooper and Carr in Chapter 10, has shown that, at present, STEM remains a relatively unknown quantity for many PSTs, and many do not see STEM being implemented during their professional placements. There needs to be deliberate linking of the

Science, Technology and Mathematics Learning Areas during initial teacher education, real opportunities for PSTs to develop their STEM teaching capacity, and assistance with them understanding the different approaches to STEM. The analysis also pointed to the need for greater emphasis on digital technology (a new area within the curriculum), and engineering (which is so far not directly included within the curriculum, although it can be incorporated within Design technologies). The problem is that many universities do not attend to Technology as a stand-alone method area at present. Some include it as an integrated Science and Technology unit where the tech component may be superficially attended to or crowded out by science because of time constraints or science-focused teacher educators. The challenge for initial teacher education is knowing how to bring in STEM-rich experiences that cover the full spectrum of STEM education as referred to by Ragloff and Guzey (2016).

Chapters 9 and 11 provided different examples of STEM enacted during initial teacher education, Chapter 9 being focused on "STEM as both pedagogy and discipline," and Ch 11 focused on bringing stakeholder groups together. In Chapter 9, Nielsen, Georgiou, Howard and Forrester draw on a range of useful pedagogies (multi-modalities, representational pedagogy [or representation construction approach], mathematics reasoning and scientific reasoning) that they believe have been effective in promoting "an integrative view of learning where the individual disciplines provide context for explorations across the fields". Giving PSTs an experience of integrative STEM through problem solving, they found, helped them see the relationships between the knowledge, skills, tools and reasoning in each subject, and enabled deeper engagement with the problem solving processes. This experience of being a learner, then metacognitive reflection on how to seek out and build connections and reasoning across the various knowledge fields, is essential for building self-efficacy, and offers valuable identity work (Calabrese Barton, Kang, Tan, O'Neill, Bautista-Guerra, & Brecklin, 2013; Lutovak & Kaasila, 2014) as teachers prepare to take STEM into their teaching.

Another experience that has this effect of identity work, and in building teaching capacity, is the mentoring partnerships that are described by Berry, McLaughlin and Cooper in Chapter 11, where PSTs mentored schoolgirls, and acted as 'bridges' or boundary spanners (Akkerman & Bakker, 2011) to facilitate interaction between the schoolgirls and STEM practitioners or technicians. This is a terrific initiative where the STEM technicians offered industry-based problems for the schoolgirls to solve, but also valuable insight into the world of work in these STEM industries and research. Seeing STEM in practice, and then realising the educational potential of these experiences, is the first step to ensuring that teachers believe in, and know how to tap into, the STEM-rich industries, companies and research institutions that can bring cutting-edge science, technology and engineering into the school as valuable contexts for learning.

4 STEM in Schools

This brings us to what is happening in schools. Half of the chapters in this book describe particular initiatives that have been implemented in schools or promote particular approaches or dispositions that may be useful in building attainment in STEM and STEM subjects.

In Chapter 2, Van Driel, Vossen, Henze and De Vries elaborate further on school-industry partnerships that focus on research and design as core practices across the STEM disciplines. As with the partnership work described by Berry et al., the partnership work described by Van Driel et al. is marked by the value of teachers collaborating with their peers and with people from STEM industries, businesses and professionals. Industry partners, who work as 'clients' in this initiative, were valuable as supervisors for student projects that focused on real world problems requiring solutions through either research or design. For schools, the reality of creating and maintaining these types of partnerships can be challenging (Education Council, 2018), so they need to be meaningful for all partners. The authors assert that collaboration between teachers from a range of schools is essential for sharing of resources, such as pedagogies that work and assessments that assess certain skills, and to share experiences.

Mathematics is the STEM subject that has been most difficult to draw into STEM in other than superficial ways. Chapters 5, 7 and 8 principally address mathematics in STEM. It is extremely powerful to have Siemon, Banks and Prasad's Chapter 5 on multiplicative thinking in this book and its necessity to successful STEM learning cannot be over-stated. Multiplicative thinking is "the capacity to recognize, represent and reason about relationships between quantities". The recent concerns about the relative decline in students' mathematics performance on international tests is linked by Siemon et al. to the high proportion of Australian students who struggle to think multiplicatively. They raise important equity issues here based on the over-representation of the lower performing students coming from lower socio-economic backgrounds. They make a strong case for taking a discipline specific approach to STEM to raise mathematical literacy, alongside the integrated approaches that tend to be promoted by STEM education advocates.

In Chapter 8, Jiang, Seah, Barkatsas, Ieong and Cheong add to this focus on improving mathematical literacy a need to attend to students' values towards mathematics. Findings from a values and preferences survey called "What I Find Important" (WIFI) showed that Grade 5/6 students from four Asian countries surveyed (Macau as compared with Chinese Mainland, Hong Kong, and Taiwan) valued achievement in mathematics most highly, followed by relevance. The research is based on the assumption that higher valuing of mathematics might lead to a higher likelihood of students continuing to study mathematics at senior levels and into tertiary studies. More important though is that they assert that poor valuing can "act against the students adopting effective learning habits and dispositions".

The construct of relevance is a value component that might be particularly attended to through STEM-rich pedagogies and contexts, especially in what Prodromou and Lavicza called the 'Big Data era' in Chapter 7. Statistics has the potential to be particularly useful for STEM as it emerges in many parts of society and in contexts that can be considered relevant for young people (such as sporting stats). Prodromou and Lavicza however state that teachers can lack an understanding of the possible role of statistics in STEM subjects. The 'Statistics inquiry cycle' is proposed where students gather information and seek meaning from that information. In groups, students completed investigations using this cycle and presented them on posters. The data showed that this approach promoted curiosity, risk-taking and negotiation of statistical meaning. Such complex problem solving, experimentation and discovery are fundamental to the STEM endeavour.

Nowhere else are curiosity, experimentation and discovery more central to the learning process than in early childhood (Campbell, Speldewinde, Howitt, & MacDonald, 2018). In Chapter 6 Gilbert and Borgerding provide a wonderful example of STEM becoming STEAM through a 'camp' situation inspired by Reggio Emilio philosophies. Adding the Arts to STEM is a controversial move, but in the context of early childhood is defensible given the strong emphasis on artistic expression at this age. Gilbert and Borgerding claim that Art enables the students to express their idea and is a conduit for STEM learning and problem-solving. Children represent their understandings through drawing, painting, modelling and construction. The artistic artifacts are constructed through the design process. Ultimately Art enables the children to "express and engage with their thoughts". A number of key learnings are presented, but most delightful and pertinent for this age group is the need for ongoing and daily reflection by the teacher, and dialogue with the children.

5 STEM at the Tertiary Level

From children to adults, in Chapter 12, McLaughlin and Kennedy focused on how to promote STEM learning at the tertiary level, but also on building the leadership capacity of academics from STEM and other disciplines to work together in cross-curricular activity in the *STEM Ecosystem Project*. This project adopts a methodological approach that exposes individuals to new forms of work and leadership roles. A 'skills ecosystem' is promoted through skills networks, and leadership training is provided for participants. One of the activities that are presented in the chapter is the 'EcoCities symposium' where people from industry, undergraduate and postgraduate students, and academics work on 'live' industry problems in urban ecology. A series of talks and opportunities to work together on solving the problems over three days promoted a 'community of practice.'

In summary, across the education spectrum, a number of different programs were presented, some funded therefore finite and potentially unsustainable, others that were more embedded. Most are locally developed and implemented so far-reaching effects are likely to be limited. However, most of the research presented here hold 'messages' for educators, such as the need to attend to values, and student perceptions and conceptualisations. Others provide valuable resources, frameworks, or ideas that have the potential to be (and in some cases have been) adopted widely, such as multiplicative thinking, the statistics inquiry cycle, 'ecosystem' and school-industry partnerships, mentoring challenges, and integrative approaches balanced with discipline-based approaches. STEM education is a dynamic field that is the responsibility of many and should be of interest to all involved in education, from childhood to adulthood. Given that we already know so much already about what works, what are the next steps for STEM?

6 A Vision for STEM Education

One of the findings from the *Our Girls – Our Future* report (Hobbs et al., 2017) was that we need to start early if we hope to influence students' attitudes towards and engagement with STEM. In this book, the full education spectrum is represented, with insights into how we can engage young children in STEM, through to school-aged children and then students at the tertiary level. A strong focus on initial teacher education provides the return loop within this education spectrum as we prepare adults to bring STEM into the lives of children and young people. Figure 13.1 represents these phases of education

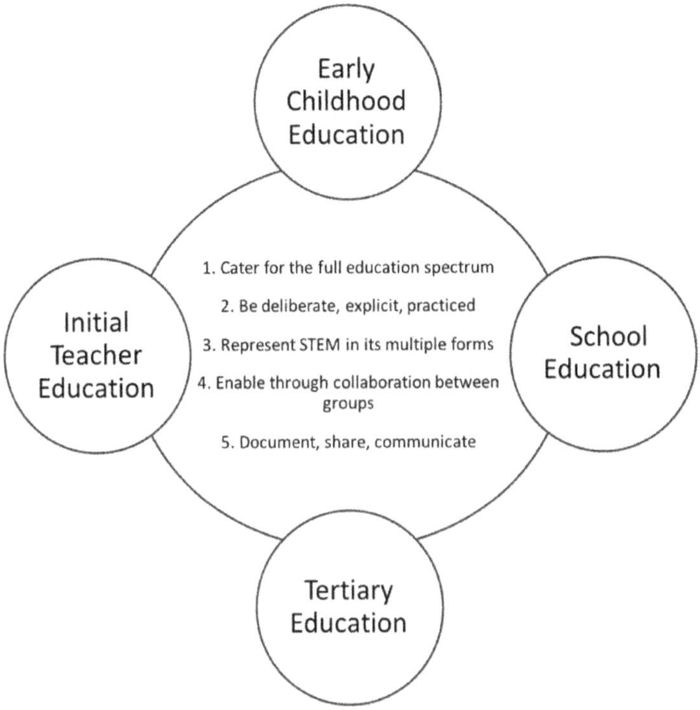

FIGURE 13.1 Principles of effective STEM education

as a cycle that is contingent on five principles. The first of these principles is to cater for this full spectrum in order to normalise STEM as an essential and attainable part of a person's complete education.

This normalization needs to be done deliberately, explicitly, and in a practiced way; the second principle. Across the chapters in this book, there are a number of examples of STEM subjects and disciplines being linked deliberately to emulate professional practice. These deliberate actions are accompanied by explicit and strong scaffolding by educators (teachers, researchers, teacher educators, lecturers) to help students understand how the subjects can be connected, and how to make those connections explicit in the case of PSTs. PSTs then need to practice teaching this way, and school and tertiary learners need to practice engaging in and with the disciplinary practices in order to develop the STEM competencies and make connections between the STEM disciplines.

How STEM is represented in education is the emphasis of the third principle. Representing STEM in its multiple forms acknowledges that there is no one penultimate way to do, learn or represent STEM. The STEM agenda arises out of political, economic and productivity concerns (Timms et al., 2018; Education Council, 2016), as was illustrated by the different discourses evident in policy documents described by Jordan in Chapter 3. These various discourses should be made explicit for students, especially in terms of how education is situated as having a critical role in this agenda. Also included within education discussions around STEM is the various ways that STEM can be represented as learning. At a philosophical and broad level, Radloff and Gusky (2016) provide a useful framework for educators considering how to approach this. At a practical level, making decisions about the degree of integration can be assisted by schemas such as those proposed by Vasquez (2015), Dugger (2010), Hobbs, Cripps Clark, and Plant (2018), and Bybee (2013). A focus on STEM pedagogies that can promote STEM thinking was illustrated by Nielsen et al. Emphasis on STEM skills, or generic skills that can be used to make connections between the STEM subjects, is a common approach in schools and universities rather than, or in addition to, disciplinary content. Siemon's warning about maintaining a focus on disciplinary learning should be heeded. Ultimately, a multi-faceted STEM strategy that uses a range of integrative and disciplinary approaches is more likely to represent the STEM endeavor more faithfully (Hobbs et al., 2018).

The fourth principle acknowledges that collaboration is a key feature of the STEM endeavor. The chapters in this book show many different ways that collaboration can be used at all levels of education, collaboration across subject groups, across industry and schools, across students, teachers, academics and industry/business/disciplinary representatives and practitioners. People are working together to: problem solve (through, for example, the STEM Ecosystem Projects involving industry, undergraduates, postgraduates and academics); to develop curriculum (through, for example, teachers working with other teachers and industry people to share resources in projects where students work with industry people as 'clients'); and develop teaching skills (through, for example, PSTs acting as boundary spanners between schoolgirls and STEM practitioners and technicians in solving industry-based problems). The focus of most of these collaborations was solving 'real world problems' using STEM knowledge and skills, which the Education Council (2018) assert should be the 'narrative' for students engaging with STEM. This collaboration has an enabling effect by opening up networks and raising an awareness and appreciation of the fact that knowledge is distributed in the education and STEM systems and is accessible if one knows where to look and who to ask. Experiencing these collaborations also can be transformative and open possibilities for the future by breaking down stereotypes (e.g., what a scientist does), improving self-efficacy (I can do this), and providing real examples of innovative educational practice that teachers, PSTs, teacher educators and other academics can realistically embrace.

The final principle is that in order to move together, there is a need to document, communicate and share such experiences as described above. Without this, each researcher, teacher, professional development provider, and policy maker are 'reinventing the wheel' every time. This principle speaks to the need for data (Education Council, 2015; Office of the Chief Scientist, 2016) to show the effects of different approaches to STEM education so that educators can be informed about the limitations, strengths and provisions associated with the different approaches. Also, educators need to document their initiatives, either through formal publication to spread the word and receive peer review, or as documents shared between teachers to illustrate how a new program is implemented in the classroom and assessed and the effects of these innovations on student learning and engagement. These efforts to document, disseminate and share resources, initiatives, and ideas have a number of purposes, they are to:

1. encourage further uptake of these innovations by other teachers;
2. assist teachers and schools to develop a comprehensive approach to STEM education;
3. reduce the effort, time and resources needed to develop new curriculum and teaching approaches;
4. build on, enhance and improve current resources;
5. encourage risk-taking but to reduce risks when blazing a new trail of STEM;
6. give direction to those new to STEM, and more advanced possibilities for educators and institutions already engaged with STEM;
7. improve student learning, attainment, dispositions, values and aspirations in relation to STEM and the STEM subjects; and
8. sustain change by embedding new teaching practices and curriculum as a normal part of the school.

Communicating about STEM requires developing a language to describe teaching approaches being adopted, the STEM knowledge and skills being emphasised and how STEM is being conceptualized. The challenge for educators is that the language available for educators is broad, varied, comprised of multiple discourses, sometimes contradictory, and sometimes unhelpful; therefore, teachers and PSTs in particular would benefit from some guidance as to what works (Education Council, 2018).

7 Conclusion

STEM education is moving beyond parsimonious versions of STEM; introducing 3D printers and doing some coding are no longer accepted as innovative responses to the STEM agenda. STEM as it is practiced is done so within disciplines, but there tends to be strong relationships between different disciplines. These varied ways of working and thinking can be represented also in schools and universities (Tytler, Osborne, Williams, Tytler, & Cripps Clark, 2008). Initial teacher education is critical to ensuring that the full education spectrum is aware of the multi-faceted nature of STEM, schools and teachers are responsible for making this happen. The authors of this book have provided some valuable examples of STEM innovation and raised some critical issues that the whole education system and policy arena need to consider when mapping out a future for STEM in education. The most important of these is how do we ensure quality STEM education across the full education spectrum? The proposed principles of effective STEM education, as arising from the research presented in this book, provide some insight into how we might, together, make this happen.

References

Akkerman, S. F., & Bakker, A. (2011). Boundary crossing and boundary objects. *Review of Educational Research, 81*(2), 132–169.

Australian Industry Group. (2016). *STEM program index 2016.* Canberra: Australian Government. Retrieved January 18, 2017, from https://www.chiefscientist.gov.au/2016/01/spi-2016-stem-programme-index-2016-2/

Bybee, R. W. (2013). *Case for STEM education: Challenges and opportunities.* Arlington, VA: National Science Teachers Association. Retrieved from http://www.ebrary.com

Calabrese Barton, A., Kang, H., Tan, E., O'Neill, T. B., Bautista-Guerra, J., & Brecklin, C. (2013). Crafting a future in science: Tracing middle school girls' identity work over time and space. *American Educational Research Journal, 50*(1), 37–75.

Campbell, C., Speldewinde, C., Howitt, C., & MacDonald, A. (2018). STEM practice in the early years. *Creative Education, 9*, 11–25. Retrieved from https://doi.org/10.4236/ce.2018.91002

Dugger, W. E. (2010, December 8–10). *Evolution of STEM in the United States.* Paper presented at the 6th Biennial Conference on Technology Education Research at Gold Coast, Australia.

Education Council. (2015). *National STEM school education strategy: A comprehensive plan for Science, Technology, Engineering and Mathematics education in Australia.* Canberra: Education Council. Retrieved July 20, 2018, from http://www.educationcouncil.edu.au/site/DefaultSite/filesystem/documents/National%20STEM%20School%20Education%20Strategy.pdf

Education Council. (2018). *Optimising STEM industry-school partnerships: Inspiring Australia's next generation.* Canberra: Education Council. Retrieved July 1, 2018, from http://www.educationcouncil.edu.au/site/DefaultSite/filesystem/documents/Reports%20and%20publications/Publications/Optimising%20STEM%20Industry-School%20Partnerships%20-%20Final%20Report.pdf

Hobbs, L., Cripps Clark, J., & Plant, B. (2018). Successful students: STEM program: Teacher learning through a multifaceted vision for STEM education. In R. Jorgensen & K. Larkin (Eds.), *STEM education in the junior secondary: The state of play* (pp. 133–168). Singapore: Springer.

Hobbs, L., Jakab, C., Millar, V., Prain, V., Redman, C., Speldewinde, C., Tytler, R., & Van Driel, J. (2017). *Girls' future – Our future: The Invergowrie foundation STEM report.* Melbourne: Invergowrie Foundation. Retrieved from http://www.invergowrie.org.au/girls-future-our-future-the-invergowrie-foundation-stem-report/

Honey, M., Pearson, G., & Schweingruber, H. (2014). *STEM integration in K-12 education: Status, prospects, and an agenda for research.* Washington, DC: The National Academies Press.

Lutovac, S., & Kaasila, R. (2014). Pre-service teachers' future-oriented mathematical identity work. *Educational Studies in Mathematics, 85*(1), 129–142.

Office of the Chief Scientist. (2016). *Australia's STEM workforce: Science, Technology, Engineering and Mathematics*. Canberra: Australian Government.

Radloff, J., & Guzey, S. (2017). Investigating changes in preservice teachers' conceptions of STEM education following video analysis and reflection. *School Science and Mathematics, 117*(3–4), 158–167. doi:10.1111/ssm.12218

Timms, M., Moyle, K., Weldon, P., & Mitchell, P. (2018). *Challenges in STEM learning in Australian schools* (Policy Insights Issue No. 7). Camberwell: ACER. Retrieved July 1, 2018, from https://research.acer.edu.au/policyinsights/7/

Tytler, R., Osborne, J., Williams, G., Tytler, K., & Cripps Clark, J. (2008). *Opening up pathways: Engagement in STEM across the primary-secondary school transition.* Canberra: Australian Government Department of Education, Employment and Workplace Relations.

Vasquez, J. (2015). Beyond the acronym. *Educational Leadership, 72*(34), 10–15.

www.ingramcontent.com/pod-product-compliance
Lightning Source LLC
Chambersburg PA
CBHW071436230926
51846CB00001B/254
9789004391383